AF572557

IEE Control Engineering Series 27
Series Editors: Prof. H. Nicholson
Prof. B.H. Swanick

PROCESS DYNAMICS ESTIMATION AND CONTROL

Previous volumes in this series:

Volume 1	Multivariable control theory J.M. Layton
Volume 2	Lift traffic analysis, design and control G.C. Barney and S.M. dos Santos
Volume 3	Transducers in digital systems G.A. Woolvet
Volume 4	Supervisory remote control systems R.E Young
Volume 5	Structure of interconnected systems H. Nicholson
Volume 6	Power system control M.J.H. Sterling
Volume 7	Feedback and multivariable systems D.H. Owens
Volume 8	A history of control engineering, 1800-1930 S. Bennett
Volume 9	Modern approaches to control system design N. Munro (Editor)
Volume 10	Control of time delay systems J.E. Marshall
Volume 11	Biological systems, modelling and control D.A. Linkens
Volume 12	Modelling of dynamical systems-1 H. Nicholson (Editor)
Volume 13	Modelling of dynamical systems-2 H. Nicholson (Editor)
Volume 14	Optimal relay and saturating control system synthesis E.P. Ryan
Volume 15	Self-tuning and adaptive control: theory and application C.J. Harris and S.A. Billings (Editors)
Volume 16	Systems Modelling and Optimisation P. Nash
Volume 17	Control in hazardous environments R.E. Young
Volume 18	Applied control theory J.R. Leigh
Volume 19	Stepping motors: a guide to modern theory and practice P.P. Acarnley
Volume 20	Design of modern control systems D.J. Bell, P.A. Cook and N. Munro (Editors)
Volume 21	Computer control of industrial processes S. Bennett and D.A. Linkens (Editors)
Volume 22	Digital signal processing N.B. Jones (Editor)
Volume 23	Robotic Technology A. Pugh (Editor)
Volume 24	Real-time computer control S. Bennett and D.A. Linkens (Editors)
Volume 25	Nonlinear system design S.A. Billings, J.O. Gray & D.H. Owens (Editors)
Volume 26	Measurement and instrumentation for control M.G. Mylroi and G. Calvert (Editors)

PROCESS DYNAMICS ESTIMATION AND CONTROL

A. JOHNSON

Peter Peregrinus Ltd on behalf of the Institution of Electrical Engineers

Published by: Peter Peregrinus Ltd., London, UK.

Cover photograph by kind permission of
Shell Nederland Raffinaderij B.V.

British Library Cataloguing in Publication Data

Johnson, A.
Process dynamics, estimation and control.—
(IEE control engineering series; 27)
1. Process control — Electronic equipment
2. Microelectronics
I. Title II. Series
670.42'7 TS5156.8

ISBN 0-86341-032-4

Printed in England by Short Run Press Ltd., Exeter

Preface

This book is written in the conviction that a new look at process control is timely. The reasons for this are threefold. First, and most importantly, because there is a lack of a comprehensive theoretical framework from which process control engineers can approach their problems; secondly, since process control is conspicuously failing to solve anything but the most simple problems using the fashionable frequency domain technique (witness the number of loops in manual control in even quite modern processes), and lastly because the importance of estimation in process control has been seriously neglected.

The reader will find here a complete and quite detailed new approach to both controller and estimator design, which has been used with success by the author and his colleagues to solve numerous industrial process control problems, in such fields as wastewater treatment, cable coating, steam production, extrusion, fermentation etc. The approach is based upon the well-understood and tested state-space techniques developed since 1960 by Kalman, Astrom, Anderson, Athans, Kwakernaak and many others.

In order to appreciate the various components of process control which have been studied in an attempt to improve upon contemporary practice, it is necessary to start with a statement of the nature of process control as meant here. Process control henceforth is the design of an algorithm for the control or the estimation of the state of a physical or chemical process. By using this very narrow definition we deliberately exclude all those (very important) topics associated with the less mathematical side of process control, such as safety, reliability, choice of instrumentation etc., although all these factors must be in the background of a worthwhile design technique.

What then are the special aspects of process control which have warranted consideration? Let us begin with the process itself. This will often contain a *time delay* owing to recycling, holdup, imperfect mixing etc. which is unfortunately neglected in most state-space approaches to process control. To remedy this, various approximations to the time delay are presented which will enable the dynamic behaviour of a large class of processes to be adequately represented for the purpose of controller or estimator design.

Secondly, the mathematical model which describes the process may consist of many state variables, as when a number of reactors or stages, for example, occur in a chemical process. This can lead to complex algorithms of large dimension for control and estimation, which, as every process control engineer knows, never work in practice. To combat this we introduce a straightforward, if somewhat crude, method of *model reduction*.

Next come the disturbances affecting the process. Traditionally these are seen in process control as deterministic (i.e. known) functions of time, whereas in practice disturbances occur randomly in time. Initial conditions, too, are never or very rarely exactly known. This leads to the adoption of a *stochastic model* to describe the process. Actually, even if our process could be described in purely deterministic terms, other factors would dictate that a stochastic model still be used. These factors stem from the fact that in process control we are not only interested in the process.

We must also be concerned with the digital computer (wherein the control or estimation algorithm resides) and with the signal reconstruction and sampling occurring in process control. Previous approaches to process control have not sufficiently emphasised the importance of including these elements in the overall analysis. It is shown here that their inclusion has two immediate consequences. First, it is necessary to use a stochastic model to account for roundoff errors, convertor noise etc. Secondly, it can be shown that a *discrete-time* dynamic model can completely describe the behaviour of the sampled-data system. How to construct this discrete-time model from the continuous-time process model is explained in some detail.

The important role of *simulation* as an aid to practical estimator and controller design should not be underestimated, and a whole chapter has been devoted to the technical details of this topic.

Finally, in this brief review of the special aspects of process control, we must mention the consideration which has been given to providing the process control engineer with *practical* estimator and controller design procedures providing numerically well-conditioned algorithms which are insensitive to small parameter variations (e.g. modelling errors etc.).

It is appropriate at this point to inform the reader what is in store for him or her. Chapter 1 contains brief reviews of the relevant topics from matrix theory, probability theory and system theory that will be required in the rest of the book. In order to achieve the stated aims the mathematics employed throughout has had to be of a fairly respectable level, consistent with that taught nowadays to engineering students at some universities and institutions.

Chapter 2 is devoted to continuous-time models. Nothing is said about mathematical modelling since so many good texts already exist on the subject. We concentrate instead on explaining how the given model can be put into a useful (i.e. linear) and compact form. Furthermore, some of the most important properties of the model are discussed; only those properties, of course, which have a direct bearing on the estimator or controller design.

Chapter 3 shows how the inclusion of the dynamics of signal sampling, signal reconstruction and digital computer can be accomplished by formulating an equivalent discrete system. This equivalent discrete system will be the starting point for our estimator or controller design techniques. Such a formulation means that the estimator or controller may be implemented in a computer directly, i.e. without any further study as to the effects of A/D or D/A conversion, finite computer wordlength etc. The properties of the equivalent discrete system are reviewed and compared with those of the (unsampled) process.

Reliable measurements may be difficult or very expensive to obtain: often it is possible to estimate the value of the variable of interest by using a filter or estimator (as we will prefer to call it) at a fraction of the cost. In Chapter 4 two such estimators are introduced: the recursive least-squares estimator and the Kalman filter. The choice of which to use is indicated by how well the behaviour of the process can be described. Sufficient conditions, in terms of the detectability and stabilisability of the process, are given to ensure the existence and the stability of the chosen estimator. We also examine various practical aspects of implementing a Kalman filter, such as what to do when the noise statistics are unknown. Akin to state variable estimation is the topic of parameter estimation or how to estimate the unknown model parameters from available measurement data. Process control engineers often have only steady-state measurements available, and so in Chapter 4 we concentrate on this type of parameter estimation.

Chapter 5 is devoted to simulation, which is an important and invaluable aid to good estimator or controller design. An initial simulation of the existing plant behaviour serves as a standard by which any future improvement can be judged. We describe in some detail one technique for simulation, the Runge-Kutta routine for the numerical integration of a set of differential equations.

The last chapter, Chapter 6, looks at controlling a process so that the process variables are held at certain fixed or setpoint values. It is well known that many problems of control may be formulated in terms of this, the basic regulator (as it is called) problem of linear optimal control. In Chapter 6 we show how the setpoints can be found by solving a static optimisation problem using the so-called Complex Method. Having found the setpoints the reader is not confronted immediately with the full solution to the setpoint control problem but led through stages which gradually built up the full-picture, starting with the deterministic setting and working towards the stochastic setpoint controller. As was the case in Chapter 4 with estimation, attention is paid not only to the theory of stochastic setpoint control, but also to how to implement such a controller in practice. For example, we examine feedforward control and also how to achieve satisfactory responses to changing setpoints. Sometimes the process control engineer has special requirements, such as there shall be no interaction between process variables; this would allow a number of single-input/single-output controllers to replace the multivariable setpoint controller. We indicate how such special requirements might be handled.

Most chapters contain worked examples, many of which are taken from the industrial control problems studied by the author during the past fifteen or so years. The examples are chosen for their didactic merits and have been stripped of all irrelevant information.

Most practising process control engineers should find this book useful and controversial. Supplemented with proofs, general reading and classroom problems it could well form the basis of a graduate level course on process control.

Acknowledgements

The suggestion to write this book came from Professor H. Nicholson. Most of the material for the book, and many of the examples, stem from various courses on control and estimation given by the author as well as from research conducted in the Department of Applied Physics of the Delft University of Technology in The Netherlands. This institution has also provided support in terms of time and secretarial facilities during the writing of this book.

I am most grateful to my two colleagues at Delft, R. M. Dekkers and W. L. de Koning for sharing with me their ideas and insights concerning process control, and the latter for reading an early draft of this book. My thanks also to my students Paul van Alphen, Manfred Voetter and Jan-Christian van der Wouden for their help with some of the figures. Finally, I thank Coby, Els and Cathy for their skill and patience with the word processor and Jackie for her conscientious proof-reading of the manuscript.

Contents

Notation and symbols

$A, B, \ldots$	uppercase letters usually denote matrices, but may denote scalars, e.g. T for time
$a, b, \ldots$	lowercase letters usually denote vectors, but may denote scalars, e.g. i, j, k
$\alpha, \beta, \ldots$	lowercase greek letters denote scalars
$a^{(i)}$	the ith element of the vector a
α_{ij}	the ijth element of the matrix A
————	denotes the beginning or end of a statement
$\mathring{x}(t)$	the derivative of the vector $x(t)$ with respect to time t, i.e. $\frac{\mathrm{d}x}{\mathrm{d}t}(t)$
$\triangleq$	equals by definition
$\equiv$	identically equal to
$\simeq$	approximately equal to
$\lvert A\rvert$	determinant of the square matrix A
$\lvert\alpha\rvert$	modulus, or absolute value, of the complex number α
$\mathrm{tr}(A)$	trace of the square matrix A
$\lambda(A)$	eigenvalue of the square matrix A
A^T	transpose of the matrix A
A^{-1}	inverse of the square matrix A
$\lVert x(\mathrm{t})\rVert$	norm of the vector $x(t)$
$\varrho(A)$	rank of the matrix A
$\mu(A)$	matrix measure of the square matrix A
$\sigma(A)$	singular value of the matrix A
$A > 0$	positive definite matrix A
$A \geqslant 0$	non-negative definite matrix A
$E\{x\}$	expectation of the vector x
$\mathrm{Re}\{\alpha\}$	real valued part of the complex number α
$\mathrm{Im}\{\alpha\}$	imaginary part of the complex number α
$\max_{\alpha}$	maximum with respect to the scalar α
$\mathrm{diag}[x]$	diagonal matrix with element $x^{(i)}$ on the ith diagonal

$U(t)$	unit step at time zero
$\delta(t)$	Dirac delta function
δ_{kl}	Kronecker function

Chapter 1

Basic definitions and results

1.1 Introduction

This chapter aims to provide the reader with a summary of the principal mathematical, statistical and system theoretic definitions and results which will be needed in the following chapters. Proofs of the results are omitted since they can be found in the references given at the end of the chapter. Particular attention has been paid to notation, which is consistent with usage in the rest of the text.

1.2 Functions

(a) *Scalar functions*
Three important functions of time are $U(t)$, $\delta(t)$ and δ_{kl}. The *unit step function*, $U(t)$ is defined by

$$U(t) \quad = \quad \begin{cases} 0 \text{ for } t \leqslant 0 \\ 1 \text{ for } t > 0 \end{cases}$$

and represents a unit step occurring at time zero. The *Dirac delta function*, $\delta(t)$ is defined by

$$\begin{cases} 0, \text{ for } t \neq 0 \\ \lim\limits_{\alpha \to 0} 1/\alpha, \text{ for } t = 0 \end{cases}$$

and has two important properties:

1 $\int_{-\infty}^{\infty} \delta(\lambda)\, d\lambda \quad = \quad 1$

2 $\int_{-\infty}^{\infty} f(\lambda)\delta(\lambda)\, d\lambda \quad = \quad f(0)$, for any function $f(t)$ that is continuous at $t = 0$.

Lastly, the *Kronecker delta function* δ_{kl} is defined by

$$\delta_{kl} = \begin{cases} 0 \text{ for } k \neq 1 \\ 1 \text{ for } k = 1 \end{cases}$$

(b) *Bounded functions*
A function $f(t)$ of time is *uniformly bounded* if $f(t) \leqslant \alpha$, some constant, for all t. A function of two variables $f(t, \tau)$ is *uniformly bounded* if $f(t, \tau) \leqslant \alpha$, where α is not dependent on either t or τ, but may be a function of $(t - \tau)$.

(c) *Matrix functions*
If one or more elements of a matrix (or vector) are functions of time, the matrix is written $A(t)$. If $A(t)$ is said to be continuous for all t, we mean that all the elements of the matrix are continuous (scalar) functions of time.

Because its elements are functions of time, it will not, in general, be possible to invert $A(t)$. We form instead the constant *Gramian (or Gram) matrix G*, defined by

$$G \triangleq \int_{t_1}^{t_2} A(t)A^{\mathrm{T}}(t)\,\mathrm{d}t$$

The Gramian matrix G is nonsingular if and only if, for any t_1 and t_2 $(> t_1)$ the elements of the rows (or columns) of $A(t)$ are a linearly independent set of time functions.

(d) *Linearisation*
A function of one or more variables is linearised by taking the first-order terms of its Taylor series expansion and disregarding the rest. Consider the scalar function $f(x, y)$. Assume that the first partial derivatives (with respect to x and y) of $f(x, y)$ be continuous and bounded. Then expanding $f(x, y)$ by means of its Taylor series about some point of reference x_r, y_r and dropping second and higher order terms, we have:

$$f(x, y) = f(x_r, y_r) + (x - x_r)\frac{\partial f}{\partial x} + (y - y_r)\frac{\partial f}{\partial y}$$

where the partial derivatives are evaluated at x_r and y_r. Note that the point of reference may itself be a function of time, i.e. $x_r(t)$ and $y_r(t)$, and that the technique of linearisation is easily extended to vector functions.

1.3 Matrix theory

(a) *Matrices and vectors*
An $(m \times n)$ matrix A is an array of elements

$$A = \begin{bmatrix} \alpha_{11} & \alpha_{12} & \cdot & \cdot & \cdot & \alpha_{1n} \\ \alpha_{21} & \alpha_{22} & \cdot & \cdot & \cdot & \alpha_{2n} \\ \cdot & \cdot & & & & \cdot \\ \cdot & \cdot & & & & \cdot \\ \cdot & \cdot & & & & \cdot \\ \alpha_{m1} & \alpha_{m2} & \cdot & \cdot & \cdot & \alpha_{mn} \end{bmatrix}$$

with each α_{ij} a real or complex number.

1 If $m = n = 1$, $A = \alpha$ is a *scalar*.
2 If $m = n$, A is a *square matrix*.
3 If $n = 1$, $A \triangleq a$ is a (column) *vector*. Sometimes the notation $a^{(i)} \triangleq \alpha_{i1}$ is used to describe the elements of a vector.
4 If

$$A = \begin{bmatrix} B & C \\ D & E \end{bmatrix}$$

where B, C, D, E are $(m_1 \times n_1)$, $(m_1 \times n_2)$, $(m_2 \times n_1)$ and $(m_2 \times n_2)$ matrices, A is a *partitioned matrix*.

(b) *Basic operations on matrices*

Equality: Given $(m \times n)A$ and $(m \times n)B$ matrices then $A = B$ if $\alpha_{ij} = \beta_{ij}$ for all i and j. A and B are said to be equal. Otherwise $A \neq B$.

Addition and subtraction: Given $(m \times n)A$ and $(m \times n)B$ matrices, matrix addition $(A + B)$ or matrix subtraction $(A - B)$ is performed by adding (or subtracting) corresponding elements.

1 $A + (B + C) = (A + B) + C$ (associative)
2 $A + B = B + A$ (commutative)

Scalar multiplication: Multiplication of any matrix by a scalar β is performed by multiplying each element of the matrix by the scalar.

Matrix multiplication: Given $(m \times r)A$ and $(r \times n)B$ matrices, the product AB is a $(m \times n)$ matrix Γ defined by $\Gamma = AB$ with

$$\gamma_{ij} = \sum_{k=1}^{r} \alpha_{ik}\beta_{kj}$$

1 $ABC = (AB)C = A(BC)$ (associative)

Transpose: Any $(m \times n)A$ matrix has a $(n \times m)$ transpose, A^T, where α_{ij} of A becomes α_{ji} of A^T.

1 $(AB)^T = B^T A^T$
2 $(A + B)^T = A^T + B^T$

Differentiation: Given any $(m \times n)A$ matrix, differentiation by a scalar is performed by differentiating each element of A, i.e. $dA/d\lambda = B$ where

$$\beta_{ij} = \frac{d\alpha_{ij}}{d\lambda}$$

Differentiation of a scalar by A, $d\lambda/dA = B$ where

$$\beta_{ij} = \frac{\partial\lambda}{\partial\alpha_{ij}}.$$

Integration is similarly defined.

Trace: The trace of a square $(n \times n)$ matrix A is the sum of all the diagonal elements, i.e.

$$\mathrm{tr}(A) = \sum_{i=1}^{n} \alpha_{ii}$$

Note $\mathrm{tr}(A)$ is a scalar.

1 $\mathrm{tr}(\kappa\{A + B\}) = \kappa\mathrm{tr}(A) + \kappa\mathrm{tr}(B)$, κ scalar, A, B square matrices
2 $\mathrm{tr}(x^T Ax) = \mathrm{tr}(Axx^T) = x^T Ax$, x vector, A square matrix
3 $\mathrm{tr}(ABC) = \mathrm{tr}(ABC)^T = \mathrm{tr}(BCA)$ with $(m \times p)$, $(p \times r)$, $(r \times m)$ matrices A, B, C.

Determinant: The determinant of a square $(n \times n)$ matrix A, denoted by $|A|$, is defined recursively by

$$A = (-1)^{i-1}\{\alpha_{i1}M_{i1} - \alpha_{i2}M_{i2} + \alpha_{i3}M_{i3} \ldots (-1)^{n-1}\alpha_{in}M_{in}\}$$

where M_{ij} is the $(n - 1 \times n - 1)$ minor of α_{ij}, which is the determinant of the matrix formed from A by striking out row i and column j.

1 $|A^T| = |A|$
2 $|AB| = |A|\,|B|$, A and B square matrices
3 $|\kappa A| = \kappa^n|A|$
4 $\begin{vmatrix} A & B \\ C & D \end{vmatrix} = |A|\,|D - CA^{-1}B|$, A and D square, and $|A| \neq 0$

(c) *Some special matrices*

Zero matrix: If every element of a matrix is zero, the matrix is a zero matrix, written 0.

Diagonal matrix: If every element of a square matrix is zero except one or more of the diagonal elements, the matrix is a diagonal matrix, written

$$\text{diag}\,(\alpha_i) \text{ where } \alpha_i \triangleq \alpha_{ii}.$$

Unit matrix: If all the diagonal elements of a diagonal matrix are ones, it is called a unit (or identity) matrix, written I.

Triangular matrix: If every element below the diagonal of a square matrix is zero, it is called an upper triangular matrix. If every element above the diagonal is zero, it is called a lower triangular matrix.

Symmetric matrix: If A is square and $A = A^{\mathrm{T}}$ then A is called a symmetric matrix.

Orthogonal matrix: If A is square and $AA^{\mathrm{T}} = I$ then A is called an orthogonal matrix.

(d) *Inverse matrices*

Singularity: Given a square matrix A, then A is called singular if $|A| = 0$, and nonsingular if $|A| \neq 0$.

Inverse: Given a square matrix A, then if and only if A is nonsingular there exists a unique matrix, denoted by A^{-1} and called the inverse of A, such that

$$A^{-1}A = AA^{-1} = I$$

There are many procedures for finding A^{-1} described in the references, and not repeated here.

1 $(A^{-1})^{\mathrm{T}} = (A^{\mathrm{T}})^{-1}$
2 $(AB)^{-1} = B^{-1}A^{-1}$, A and B square, nonsingular matrices.
3 $(A + BCD)^{-1} = A^{-1} - A^{-1}B(C^{-1} + DA^{-1}B)^{-1}DA^{-1}$, A and C square, nonsingular matrices (the matrix inversion lemma).
4 $(A + BCD)^{-1}BC = A^{-1}B(C^{-1} + DA^{-1}B)^{-1}$, A and C square, nonsingular.

(e) *Eigenvalues*

Characteristic polynomial: Given a $(n \times n)$ square matrix A and a scalar λ, the nth order polynomial $|\lambda I - A|$ is called the characteristic polynomial of A

1 Let $|\lambda I - A| = \lambda^n + \gamma_1\lambda^{n-1} + \ldots + \gamma_n$.
Then $A^n + \gamma_1 A^{n-1} + \ldots + \gamma_n I = 0$ (Cayley-Hamilton)

Eigenvalues: The n roots of the characteristic polynomial are the eigenvalues, denoted by $\lambda_1, \lambda_2, \ldots, \lambda_n$ or $\lambda_i(A)$ of the square matrix A. An eigenvalue is called distinct if it is not a repeated root.

1 $\sum_{i=1}^{n} \lambda_i = \mathrm{tr}(A)$

2 $\lambda_1 \lambda_2 \lambda_3 \ldots \lambda_n = |A|$
3 If $|A| \neq 0$, $\lambda_i(A^{-1}) = \lambda_i^{-1}(A), \quad i = 1, 2, \ldots, n$
4 $\lambda_i(e^{At}) = e^{\lambda_i(A)t}, \quad i = 1, 2, \ldots, n$
5 $\lambda_i(A^T) = \lambda_i(A), \quad i = 1, 2, \ldots, n$
6 $\lambda_i(A + \gamma I) = \lambda_i(A) + \gamma, \quad i = 1, 2, \ldots, n$
7 $|A| = 0 \Leftrightarrow \lambda_i(A) = 0$ for some i

Spectrum: The set of all the distinct eigenvalues of the square matrix A is called the spectrum of A.

Spectral radius: The maximum absolute value of the eigenvalues of the square matrix A is the spectral radius of A, i.e.

$$\max_i (|\lambda_i(A)|).$$

Singular values: Given a $(m \times n)$ matrix A, with $n \geqslant m$, then the singular values, $\sigma_i(A)$, of A are the m non-negative values given by

$$\sigma_i(A) = \{\lambda_i(AA^T)\}^{1/2}, \qquad i = 1, 2, \ldots, m$$

Matrix measure: Given a real square matrix A. Then the matrix measure, $\mu(A)$, is given by

$$\mu(A) = \max_i \tfrac{1}{2}\lambda_i(A + A^T)$$

(f) *Functions of a matrix*

Powers: Given a square matrix A and some scalar $m > 0$, we define A^m as the product of m matrices A. For negative m, let $m = -n$ then $A^m = (A^{-1})^n$.

Exponential: The matrix exponential, e^{At}, is given by

$$e^{At} = I + At + \frac{1}{2!} A^2 t^2 + \ldots$$

1 $e^{-At} = (e^{At})^{-1}$

(g) *Rank of a matrix*
The rank of a $(m \times n)$ matrix A, denoted by $\varrho(A)$, is the number of nonzero

singular values of A. It is greater than, or equal to, the number of nonzero eigenvalues of A, equality holding when A is symmetric.

1 $\varrho(A) = 0 \Leftrightarrow A = 0$
2 $\varrho(A) = n \Leftrightarrow |A| \neq 0$, A square
3 $\varrho(AB) = \varrho(A)$, B square, nonsingular
4 $\varrho(AB) \leqslant \min(\varrho(A), \varrho(B))$
5 $\varrho(A + B) \leqslant \varrho(A) + \varrho(B)$
6 $\varrho(A^T A) = \varrho(A) = \varrho(AA^T) = \varrho(A^T)$

(h) *Norms of vectors and matrices*

Vector norm: The (Euclidean) vector norm, denoted by $\|x\|$ is defined as

$$\|x\| = \{x^T x\}^{1/2}$$

Spectral norm: The matrix norm corresponding to the vector norm just given is the spectral norm. The spectral norm of an $(m \times n)$ matrix A, denoted by $\|A\|$, is defined as the largest singular value of A.

1 $\|A\| = 0 \Leftrightarrow A = 0$
2 $\|Ax\| \leqslant \|A\| \, \|x\|$
3 $\|A + B\| \leqslant \|A\| + \|B\|$
4 $\|AB\| \leqslant \|A\| \, \|B\|$

(i) *Positive and non-negative definite, symmetric matrices*

Positive definite matrix: Given a square, symmetric matrix A, then A is positive definite, denoted by $A > 0$, if for all nonzero vectors x the scalar quantity $x^T Ax > 0$.

Non-negative definite matrix: Given a square, symmetric matrix A, then A is non-negative definite (sometimes called semipositive definite), denoted by $A \geqslant 0$, if for all nonzero vectors x, $x^T Ax \geqslant 0$.

1 $A > 0 \Leftrightarrow \lambda_i(A) > 0$ for all i
2 $A \geqslant 0 \Leftrightarrow \lambda_i(A) \geqslant 0$ for all i
3 $A > 0 \Leftrightarrow |A| \neq 0$
4 $A > 0 \Rightarrow A^{-1} > 0$
5 $A > 0 \Rightarrow \alpha_{ii} > 0$ where α_{ii} are the diagonal elements of A
6 $A \geqslant 0 \Rightarrow \alpha_{ii} \geqslant 0$ where α_{ii} are the diagonal elements of A

Symmetric square root: Given a square symmetric matrix A, then there exists a symmetric square root, having the same dimensions as A and denoted by $A^{1/2}$, such that $A^{1/2}(A^{1/2})^T = A^{1/2} A^{1/2} = A$. If, moreover, $A > 0$ then $A^{1/2}$ is unique.

1 $A \geqslant 0 \Rightarrow A^{1/2} \geqslant 0$
2 $A > 0 \Rightarrow A^{1/2} > 0$

1.4 Probability theory

(a) *Random variables*

Suppose n random variables are collected together in the $(n \times 1)$ vector x. All n variables take real values. If these values can be collected into a finite (or countable) set of distinct values, then x will be called a discrete random variable vector, otherwise a continuous random variable vector. For brevity, where no misconceptions can arise, x will sometimes just be called a random variable.

The properties of random variables that interest us will be described using the *expectation operator*, E. The expectation operator acting on a scalar continuous random variable ξ is defined by

$$E\{\xi\} = \int_{-\infty}^{\infty} \lambda p(\lambda)\, \mathrm{d}\lambda$$

where $p(\lambda)$ is the probability density associated with ξ. If ξ is a discrete scalar random variable, taking the values λ_i with probability p_i then

$$E\{\xi\} = \sum_{i=0}^{\infty} \lambda_i p_i$$

When random variable vectors are involved the expectation operator acting on x, $E\{x\}$, is taken to mean that it acts on each of the component random variables as above. Obviously, it must be assumed that both the integral and the series in the definitions are absolutely convergent, otherwise $E\{x\}$ is not defined.

(b) *Mean and covariance of random variables*

The mean of x, sometimes written $\bar{x}$, is defined as $E\{x\}$. The covariance of x, sometimes written cov x, is the $(n \times n)$ matrix given by

$$E\{(x - \bar{x})(x - \bar{x})^{\mathrm{T}}\}$$

If x is a scalar random variable, the covariance is replaced by the scalar quantity $E\{(x - \bar{x})^2\}$, called the variance. Note that the variance is always non-negative, and the covariance always a symmetric matrix which is non-negative definite (see previous section). Only these first two moments, the mean and covariance, are needed in this book.

(c) *Properties of the expectation operator*

Some of the most useful properties of $E\{x\}$ are now listed. Here A and B are (nonrandom) matrices and c is a fixed (nonrandom) vector; y is another random variable.

1 $E\{c\} = c$
2 $E\{Ax + By\} = AE\{x\} + BE\{y\}$
3 $E\{E\{x\}\} = \bar{x}$

(d) *Correlation and independence*

Let there be two random variable vectors, x and y. The covariance of x and y is the $(n \times n)$ matrix given by

$$E\{(x - \bar{x})(y - \bar{y})^{\mathrm{T}}\}$$

and if this covariance of x and y is zero then x and y are said to be *uncorrelated.* If, on the other hand, the pair x and y are *independent* then

$$E\{xy^{\mathrm{T}}\} = E\{x\}E\{y^{\mathrm{T}}\}$$

Since

$$E\{(x - \bar{x})(y - \bar{y})^{\mathrm{T}}\} = E\{xy^{\mathrm{T}}\} - \bar{x}\bar{y}^{\mathrm{T}}$$

it is clear that for any pair of random variables, independence implies that the pair of random variables are uncorrelated. In general, however, it is not true to state that because two variables are uncorrelated, they are independent. Zero mean variables that are uncorrelated are, of course, independent.

(e) *Normal (or Gaussian) random variables*

Suppose the random variable vector x has a nonsingular covariance, i.e..

$$E\{(x - \bar{x})(x - \bar{x})^{\mathrm{T}}\} = \Sigma$$

where $|\Sigma| \neq 0$. Then x is said to be normal or Gaussian if its probability density has the form

$$p(y) = \frac{1}{(2\pi)^{n/2}} \frac{1}{|\Sigma|^{1/2}} \exp\{-\tfrac{1}{2}(y - \bar{x})^{\mathrm{T}}\Sigma^{-1}(y - \bar{x})\}$$

The statistical properties of a normal random variable are completely determined by its mean and covariance; sometimes we write x is $N(\bar{x}, \Sigma)$ meaning normally distributed with mean $\bar{x}$ and covariance Σ. If two normal variables are uncorrelated, it can be proved that they are independent.

(f) *Random processes*

Random (or stochastic) processes are essentially time dependent random variables. A continuous random process $x(t)$ is an $(n \times 1)$ vector of scalar random processes each exhibiting a continuous dependence on time t. A discrete random process, x_k, is sometimes denoted as a sequence of random vectors $\{x_k\}$ obtained at discrete instants, $k = 0, 1, 2 \ldots$ of time. All that has been said of random variables carries over to random processes, and so will not be repeated here. Note that means and covariances become (in general) functions of time. For example, the covariance of a discrete random process is given by

$$E\{(x_k - \bar{x}_k)(x_l - \bar{x}_l)^{\mathrm{T}}\}$$

and so depends on k and l. A *stationary random process* is one where the mean is constant and the covariance depends upon the difference $(k - l)$ only.

(g) *White noise processes*
A very special random process is called white noise. We define continuous white noise as a continuous random process $x(t)$ such that

$$E\{x(t)\} = 0$$

$$E\{x(t)x^{\mathrm{T}}(s)\} = N(t)\delta(t - s)$$

where $\delta(.)$ is the Dirac delta function, and discrete white noise as a discrete random process $\{x_k\}$ such that

$$E\{x_k\} = 0$$

$$E\{x_k x_l^{\mathrm{T}}\} = W_k \delta_{kl}$$

where δ_{kl} is the Kronecker delta function. The $(n \times n)$ matrices $N(t)$ and W_k are non-negative definite, symmetric matrices for all t and k, respectively.

1.5 State equation

(a) *Continuous transition matrix* $\Phi(t, \tau)$
Let $A(t)$ be continuous for all t, then $\Phi(t, \tau)$ is defined as the solution of

$$\frac{\mathrm{d}}{\mathrm{d}t}\Phi(t, \tau) = A(t)\Phi(t, \tau)$$

$$\Phi(\tau, \tau) = I$$

If $A(t) = A$,

$$\Phi(t, \tau) = \mathrm{e}^{A(t-\tau)}$$

Properties of the transition matrix are:

1 $\Phi(t, \tau) = \Phi(t, \lambda)\Phi(\lambda, \tau)$
2 $\Phi^{-1}(t, \tau) = \Phi(\tau, t)$
3 $|\Phi(t, \tau)| \neq 0$
4 $\Phi(t, \tau) = \theta(t)\phi^{-1}(\tau)$

for all t, λ and τ.

(b) *Finite-dimensional continuous state equation*
This is the differential equation

$$\dot{x}(t) = A(t)x(t) + B(t)u(t)$$

$$x(t_0) = x_0, \text{ constant}$$

If $A(t)$ is continuous, the unique solution of the above is

$$x(t) = \Phi(t, t_0)x_0 + \int_{t_0}^{t} \Phi(t, \tau)B(\tau)u(\tau)\,\mathrm{d}\tau$$

where $\Phi(t, \tau)$ is the transition matrix of $A(t)$.

(c) *State equation with time delay*
The differential-difference equation

$$\dot{x}(t) = A_0(t)x(t) + A_1(t)x(t - T) + B(t)u(t)$$

$$x(t_0) = x_0, \text{ constant}$$

has the unique solution, if $A_0(t)$ is continuous, given by

$$x(t) = \Phi(t, t_0)x_0 + \int_{t_0}^{t} \Phi(t, \tau)[A_1(\tau)x(\tau - T) + B(\tau)u(\tau)]\, d\tau$$

where $\Phi(t, \tau)$ is the transition matrix of $A_0(t)$.

(d) *The discrete transition matrix* $\Phi_{k,l}$
This is defined as the solution of the difference equation

$$\Phi_{k+1,l} = A_k\Phi_{k,l}, \quad k \geqslant l$$

$$\Phi_{l,l} = I$$

Obviously, from this definition $\Phi_{k,l} = A_{k-1}A_{k-2} \ldots A_l$ for $k \geqslant l + 1$. If $A_k = A$ then $\Phi_{k,l} = A^{k-l}$

1 $\Phi_{k,l} = \Phi_{k,m}\Phi_{m,l}$
2 $|\Phi_{k,l}| = |A_{k-1}|\,|A_{k-2}| \ldots |A_l|$
3 If $|\Phi_{k,l}| \neq 0$:

$$\Phi_{k,l}^{-1} = \Phi_{l,k}$$

4 $\Phi_{k,l} = \theta_k\phi_l^{-1}$ if $|\Phi_{k,l}| \neq 0$

(e) *Finite-dimensional discrete state equation*
This is the difference equation

$$x_{k+1} = A_kx_k + B_ku_k$$

$$x_0 = \text{constant}$$

The unique solution is, for $k \geqslant 1$

$$x_k = \Phi_{k,0}x_0 + \sum_{i=0}^{k-1} \Phi_{k,i+1}B_iu_i$$

where $\Phi_{k,l}$ is the transition matrix of A_k.

(f) *State transformations*
Let $\tilde{x}(t) = Sx(t)$, where S is nonsingular. Then

$$\dot{\tilde{x}}(t) = SAS^{-1}\tilde{x}(t) + SBu(t)$$

$$\tilde{x}(0) = Sx_0$$

is said to be *equivalent* to the equation

$$\mathring{x}(t) = Ax(t) + Bu(t)$$

$$x(0) = x_0$$

The matrices A and SAS^{-1} are said to be *similar* and the transformation is a system similarity transformation.

1 $S\Phi(t, \tau)S^{-1}$ is the transition matrix of SAS^{-1}, where $\Phi(t, \tau)$ is the transition matrix of A
2 $\lambda_i(SAS^{-1}) = \lambda_i(A), \quad i = 1, 2, \ldots, n$
3 (SAS^{-1}, SB) controllable if and only if (A, B) controllable
4 (CS^{-1}, SAS^{-1}) observable if and only if (C, A) observable

(g) *State-space model*

If the state equation is supplemented by an (algebraic) output equation we have a state-space model. The finite-dimensional continuous linear state-space model is

$$\mathring{x}(t) = A(t)x(t) + B(t)u(t)$$

$$x(t_0) = x_0, \text{ constant}$$

$$y(t) = C(t)x(t)$$

while its discrete counterpart is

$$x_{k+1} = A_k x_k + B_k u_k$$

$$x_0 = \text{constant}$$

$$y_k = C_k x_k$$

If $A(t)$, $B(t)$ and $C(t)$ are constant matrices and if $x_0 = 0$ the Laplace transformation applied to the state-space model yields

$$y(s) = C(sI - A)^{-1}Bu(s)$$

For scalar $u(t)$ and $y(t)$ the above expression may be written as a rational function of s, the Laplace parameter:

$$y(s) = \frac{(s - z_1)(s - z_2) \ldots (s - z_k)}{(s - p_1)(s - p_2) \ldots (s - p_l)} \cdot u(s)$$

where $k < l$. The complex numbers $z_1, z_2 \ldots z_k$ are known as the zeros and the complex numbers $p_1, p_2 \ldots p_l$ the poles of the model.

References

AYRES, F. (1974): 'Matrices' (McGraw-Hill)
CHEN, C. T. (1970): 'Introduction to linear system theory' (Holt, Rinehart and Winston)
PAPOULIS, A. (1965): 'Probability, random variables and stochastic processes' (McGraw-Hill)
ROSENBROCK, H. H. (1970): 'State-space and multivariable theory' (Nelson)
WIBERG, D. M. (1971): 'State space and linear systems' (McGraw-Hill)

Chapter 2

Continuous-time process models

2.1 Introduction

A mathematical model is a prerequisite for good controller or estimator design. Not only will simulation, prior to the implementation of a controller or estimator, require a mathematical model, but the controller or estimator will need information concerning the process dynamics from the mathematical model as well as from measurements in order to function properly. Some excellent texts (Himmelblau and Bischoff, 1968; Douglas, 1972; Friedly, 1972; Fasol and Jorgl, 1980; Roffel and Rijnsdorp, 1982) are available which explain the construction of mathematical models of dynamical processes, therefore we need say nothing about that here. Rather, the aim of this chapter is twofold: to classify the models and to investigate their properties which have relevance for process control and/or estimation. It will be assumed throughout that the parameters of the mathematical model can be estimated from process measurements (see Chapter 4) or assigned approximately correct numerical values. Furthermore, the mathematical model is assumed to describe not only the dynamic behaviour of the plant or process under investigation, but also the dynamics of associated actuators and sensors. The operations of signal sampling and reconstruction are not modelled – this will be done in Chapter 3.

The mathematical models describing the process dynamics usually contain variables which are continuous functions of time, since they are derived from the continuity equation and/or one or more of the laws of conservation of mass, energy or momentum. We will call them continuous-time (or sometimes just continuous) models. Some situations, such as when measurements from an instrument are available only at certain discrete instants of time, cannot be described by continuous models. We will consider such cases in Chapter 3 when discrete-time models are introduced. Actually, it will be necessary to transform all our continuous models into discrete models to account for sampling and reconstruction of signals but this need not concern us at the moment.

In addition to modelling the process, we must model the disturbances which affect it. The most realistic way to do this is to include various stochastic components in the mathematical model. Moreover the inclusion of these stochastic processes in the process description does not substantially increase the mathematical effort required to solve control or estimation problems in Chapters 4 and 6.

The model classification indicated in this chapter is based upon the form of the step response. The classification is useful when checking simulations (Chapter 5) and would enable the experienced process control engineer to model complex large-scale processes from a number of 'dynamical building blocks'. As mathematical modelling is a very time consuming activity, the savings here can be substantial. Moreover, the model classification should aid our model reduction. By this we mean obtaining the smallest dimension model – in other words the least number of equations – such that the process dynamics is adequately represented by the model. Working with a small dimension model is essential for a practical controller or estimator design.

The second half of this chapter is devoted to a study of those properties of dynamical processes which are important for controller or estimator design. Since it has already been indicated that our continuous model will be transformed later to a discrete model, the reader may question the necessity of such a study. It turns out that it is far more difficult to relate the properties of the discrete model to the actual process. Moreover, the dynamical properties of the process may be masked by those of the samplers.

Lastly, we mention the most serious shortcomings of this chapter. An important class of mathematical models (distributed parameter models) have been omitted, although one special case which is particularly important in process control (the time-delay model) is included. Model reduction techniques are not described except for the heuristic approach of least-squares response fitting. Finally, some less important dynamical properties have been omitted. In spite of those shortcomings, the chapter should be useful in a majority of practical situations such as those encountered by the process control engineer.

2.2 Model classification

2.2.1 *Finite-dimensional models*

When the dynamic behaviour of a process may be adequately described by a finite number of ordinary differential equations we speak of a finite-dimensional (or sometimes of a lumped parameter) model. Models involving partial differential equations which reflect the spatial as well as the temporal dependence of the model state variables are called infinite-dimensional (or distributed parameter) models. A worthwhile study of infinite-dimensional models is beyond the scope of this book, although we note that the technique of spatial discretisation (see, for example, Roffel and Rijnsdorp, 1982) enables some distributed parameter models to be satisfactorily approximated by finite-dimensional models. In Section 2.2.4 a special, but extremely important subclass of infinite-dimensional models exhibiting time delay are described. Thus the applicability of the models to be investigated is quite wide.

To make matters more precise, we have the following definition:

Let all variables and functions be real valued. Then a finite-dimensional continuous model consists of the (vector) state differential equation*

$$\mathring{x}(t) = f[x(t), u(t), d(t), t] \tag{2.1}$$

$$x(t_0) = x_0 \tag{2.2}$$

and the algebraic (vector) output equation

$$y(t) = g[x(t), u(t), d(t), t] \tag{2.3}$$

with $x(t) \in R^n$, $y(t) \in R^m$, $u(t) \in R^r$ and $d(t) \in R^l$

The n-dimensional vector $x(t)$ is the *state* of the model and its initial value is the constant, x_0. Some disturbances may be conveniently modelled through x_0 (see Section 2.2.2). The r-dimensional vector $u(t)$ contains only the manipulated (i.e. control) variables, sometimes called the input. There must therefore be r actuators. All those unmanipulated independent variables, including some of the disturbances affecting the process, are collected together in the l-dimensional vector, $d(t)$. Finally, $y(t)$ is the m-dimensional vector of the measured variables, sometimes called the output. There must therefore be m sensors. It is not usually necessary to specify in addition the unmeasured output variables.

Some common terminology must be explained. Since n, the number of differential equations in the model (and called the model order), is a finite number, we call the model a finite-dimensional model. The variables are all continuous functions of time, so the model is called a continuous finite-dimensional model. If either $r \neq 1$ or $m \neq 1$ we talk of a multivariable model; otherwise it is s.i.s.o. (single-input/single-output).

Since any nth-order ordinary differential equation may be written as n first-order differential equations, the formulation given is quite general. It will represent a large number of mathematical models that we wish to utilise.

2.2.2 *Disturbances*

The model of the previous section included disturbances, $d(t)$, which were defined as the independent and unmanipulated variables. Some examples of disturbances are:

the ambient temperature;
an uncontrolled flow to a reactor vessel;
the charge to a batch reactor which changes depending on product demand.

* An explanation of the notation is given in the beginning of this book.

With this terminology it is possible, it should be noted, to measure a disturbance and this can be put to advantage in controller design (see Chapter 6).

To describe disturbances in a realistic way requires that $d(t)$, in general, be a random process or random variable (see Chapter 1). Thus disturbances which are best described as initial conditions, x_0, on the process model of the last section, are random variables (with known or estimated mean, covariance and distribution) while all other disturbances are random processes. These random processes consist (we postulate) of a constant *bias component*, d_r and a *stochastic component* having a zero mean. Sometimes this disturbance model will prove too crude (although it is always a good starting point) and we will need to superimpose the stochastic component on a dynamical (i.e. nonconstant) carrier: usually, a linear difference equation forced by white noise is sufficient for the purpose. An example can be found in Chapter 6.

A consequence of $d(t)$ being random is that the process model of the last section is mathematically unsound. The differential equations need to be replaced by stochastic integral equations; however, we will sacrifice mathematical rigour here in the knowledge that a more detailed analysis confirms our results and is likely to be of little interest to the average process control engineer.

2.2.3 *Linearisation*

The derivation of the finite-dimensional process model will usually have involved a compromise between the accuracy of representing variables and the complexity of the equations involved. A further simplification is now proposed, namely that a linear model be constructed from the nonlinear finite-dimensional process model. The nonlinear model will still be used for both process optimisation (Chapter 6) and simulation (Chapter 5). The benefits of linearisation, such as the insight it brings in fundamental dynamic properties and the access to well-developed controller and estimator design procedures, far outweigh the time spent on linearisation and the loss of accuracy. Note that a linear model can be chosen to give an arbitrarily accurate approximation about a point to the true behaviour (Desoer and Vidyasagar, 1975), and moreover, good (i.e. robust) controllers and estimators must be able to cope with small modelling inaccuracies.

If measurements of the process variables are available and are to be put to use to determine unknown model parameters it may be wise to do this before linearisation. There is often the possibility (Chapter 4) of using steady-state data to obtain parameter values of nonlinear models, while linear models would necessitate dynamic tests.

Linearisation proceeds as follows. Assume that the disturbances affecting the finite dimensional model described by eqns. 2.1 to 2.3 have no stochastic components. We will come back to this simplification later. Then letting some known reference (or nominal) input $u_r(t)$ and initial condition x_{r0} be applied to the system during $t_0 \leqslant t \leqslant t_1$

$$\mathring{x}_r(t) = f[x_r(t), u_r(t), d_r, t] \tag{2.4}$$

$$x_r(t_0) = x_{r0} \quad (2.5)$$

$$y_r(t) = g[x_r(t), u_r(t), d_r, t] \quad (2.6)$$

Here d_r is the constant bias component of the disturbance. Now define *perturbation variables* (or deviation variables) to be the difference between the actual and reference trajectories

$$\Delta x(t) \triangleq x(t) - x_r(t) \quad (2.7)$$

$$\Delta u(t) \triangleq u(t) - u_r(t) \quad (2.8)$$

$$\Delta y(t) \triangleq y(t) - y_r(t) \quad (2.9)$$

and linearise (Chapter 1) eqns. 2.1 to 2.3 to yield the model

$$\Delta \mathring{x}(t) = A(t)\Delta x(t) + B(t)\Delta u(t) + w(t) \quad (2.10)$$

$$\Delta x(t_0) = x_0 - x_{r0} \quad (2.11)$$

$$\Delta y(t) = C(t)\Delta x(t) + D(t)\Delta u(t) \quad (2.12)$$

which is valid for $t_0 \leqslant t \leqslant t_1$. The term $w(t)$, given by

$$w(t) = f[x_r(t), u_r(t), d_r, t] - \mathring{x}_r(t) \quad (2.13)$$

will be zero in view of eqn. 2.4. However, it does show how any modelling errors or inaccuracies could propagate through the linearisation procedure.

We now impose the important restriction that $D(t) = 0$, in other words that the process under consideration contains no direct link between input and output. The restriction is necessary for our estimator and controller design techniques presented in later chapters. Of course any pure gain present in the process and modelled by a nonzero $D(t)$ can be approximated by a very fast state variable which transmits input information almost instantaneously to the output, $y(t)$. Moreover we might be able to define a new output variable so that the direct link in the output equation vanishes.

With $D(t) \equiv 0$ and when $w(t) = 0$, the model is called (see also Chapter 1) the deterministic finite-dimensional linear *state-space model*:

$$\mathring{x}(t) = A(t)x(t) + B(t)u(t) \quad (2.14a)$$

$$x(t_0) = x_0 \quad (2.14b)$$

$$y(t) = C(t)x(t) \quad (2.14c)$$

where $x(t)$, $u(t)$ and $y(t)$ are now perturbation variables; in fact this is often the nomenclature adopted (although frequently unstated) in the literature. It should be pointed out that even when eqns. 2.4 and 2.6 do not explicitly depend on time, the matrices $A(t)$, $B(t)$ etc. may still be time varying. If, however, the input is constant (i.e. u_r) and x_r is chosen such that

$$0 = f(x_r, u_r, d_r, t) \quad (2.15)$$

then the matrices A, B, etc. are constant, and the linear state-spae model is said to be *time invariant*.

The accuracy of the approximation of the linear model to the true behaviour of the state and output of the process depends upon:

the magnitude of second and higher order terms of the Taylor series discarded during linearisation;
the choice of reference trajectories $u_r(t)$ and $x_r(t)$;
so far as circumstances permit, the choice of actuators and sensors (i.e. the choice of $u(t)$ and $y(t)$). See Example 2.1.

As a check on the accuracy it is strongly recommended that the responses, both to nonzero initial state values and to step forcing functions, of the linear model and the nonlinear model are compared by means of simulation (Chapter 5).

The matrices $A(t)$, $B(t)$ etc. contain (deterministic) parameter values which depend, in general, upon the disturbances d_r. Because exact values of the disturbances are usually unknown, the parameters of the state-space model are inherently inaccurate. However, it is reassuring, on the other hand, that in studying the properties of the linear state-space model the effect of the disturbances has to some extent been included. What then, of the neglected stochastic component of the disturbance? Heuristically, we argue that if these components had been included the matrices $A(t)$, $B(t)$ etc. would then contain *random parameters* as elements. So properties of the linear state-space model with deterministic parameters are 'mean' properties. An approach often adopted is to partially account for the stochastic component of the disturbance by including additive white noise (Chapter 1):

$$\mathring{x}(t) = A(t)x(t) + B(t)u(t) + w(t) \tag{2.16a}$$

$$x(t_0) = x_0 \tag{2.16b}$$

$$y(t) = C(t)x(t) + v(t) \tag{2.16c}$$

The initial condition is usually also taken to be a random variable. The initial condition x_0 and the processes $\{v(t)\}$ and $\{w(t)\}$ are assumed to be mutually uncorrelated, with known means and covariances

$$E\{x_0\} = \bar{x}_0 \tag{2.17a}$$

$$E\{(x_0 - \bar{x}_0)(x_0 - \bar{x}_0)^{\mathrm{T}}\} = P, \quad P \geqslant 0 \tag{2.17b}$$

$$E\{w(t)\} = 0 \tag{2.17c}$$

$$E\{w(t)w^{\mathrm{T}}(s)\} = W(t)\delta(t-s), \quad W(t) \geqslant 0 \tag{2.17d}$$

$$E\{v(t)\} = 0 \tag{2.17e}$$

$$E\{v(t)v^{\mathrm{T}}(s)\} = V(t)\delta(t-s), \quad V(t) \geqslant 0 \tag{2.17f}$$

$$E\{w(t)v^{\mathrm{T}}(s)\} = S(t)\delta(t-s), \quad S(t) \geqslant 0 \tag{2.17g}$$

Actually, eqn. 2.16*a* should be rigorously defined in terms of a stochastic integral equation forced by a Wiener process, in accordance with our remark in the previous section. Note also that the properties that will interest us (Section 2.3) of the stochastic state-space model, eqns. 2.16, are the same as those of the deterministic state-space model, eqns. 2.14.

Example 2.1

A mathematical model of the process shown in Fig. 2.1 is the following:

$$\varrho A \frac{\mathrm{d}h(t)}{\mathrm{d}t} = F_i(t) - k \cdot c(t)\sqrt{h(t)}$$

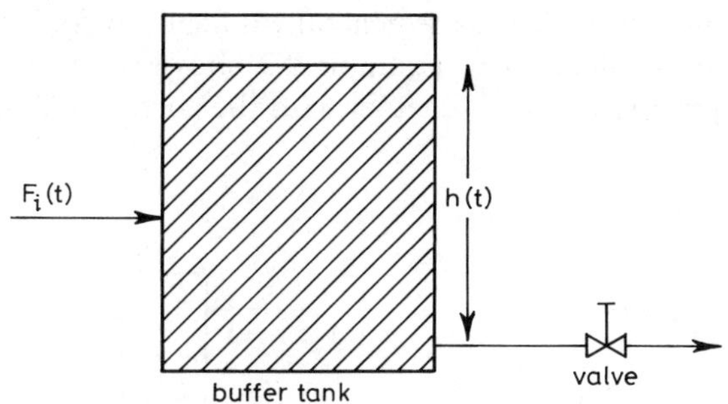

Fig. 2.1 *Buffer tank process of Example 2.1*

where

ϱ	is the fluid density, kg/m^3
A	is the cross-sectional area of the tank, m^2
h	is the height of the fluid, m
$F_i(t)$	the unregulated inflow, kg/s
k	a physical constant, kg/m$^{3/2}$s
$c(t)$	the variable outflow pipe cross-sectional area, m^2

Here, all constants and variables are scalars.

The unregulated inflow is modelled as consisting of a constant bias $\bar{F}$ and a zero-mean random fluctuation $f(t)$. We choose to control the process via $c(t)$. Linearisation about the reference point h_r, which is chosen to be a steady-state of the process corresponding to some fixed value c_r of the valve cross-sectional area, i.e.

$$\bar{F} = kc_r\sqrt{h_r}$$

gives

$$\frac{\mathrm{d}\Delta h(t)}{\mathrm{d}t} = \frac{-\bar{F}}{2\varrho A h_r}\Delta h(t) - \frac{\bar{F}}{\varrho A c_r}\Delta c(t) + \frac{f(t)}{\varrho A}$$

Note how this equation, together with a suitable initial condition, $\Delta h(t_0)$, is a stochastic linear state equation. If the height of fluid in the buffer tank were measured, $y(t) = h(t)$ would be the output equation, completing the model. Note how the constant component of $F_i(t)$, which we may only be able to roughly estimate, appears in the model parameters.

2.2.4 *Time-delay models*

Certain processes, such as those involving considerable lengths of piping or cascaded mixers, dead-zones etc. give rise to time delays (sometimes called dead times, transportation lags or time lags). Some sensors also have a time delay. Processes containing time delays are generally difficult to control or to estimate and need special consideration. Example 2.2 shows some typical situations encountered in process control, together with the corresponding mathematical models.

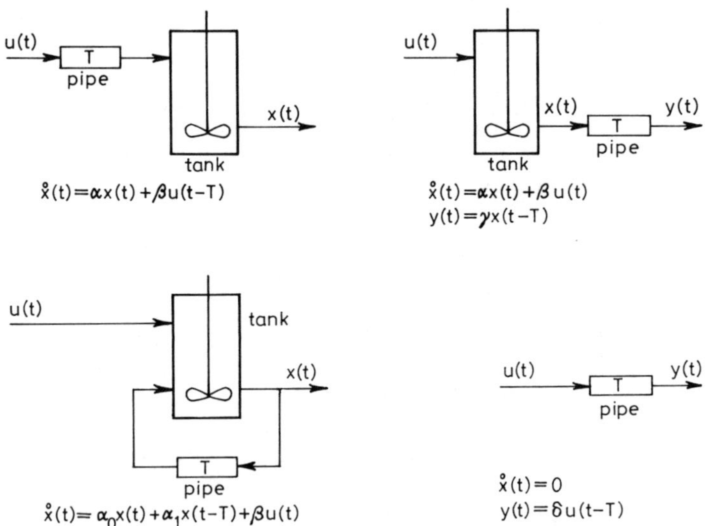

Fig. 2.2 *Four configurations of piping and perfectly-mixed vessels illustrating the different types of time-delay models*

Example 2.2

See Fig. 2.2 for four configurations of piping and perfectly-mixed vessels, illustrating the different types of time-delay models, with the following equations:

$$(a)\quad \mathring{x}(t) = \alpha x(t) + \beta u(t - T)$$

(*b*) $\mathring{x}(t) = \alpha x(t) + \beta u(t)$

$y(t) = \gamma x(t - T)$

(*c*) $\mathring{x}(t) = \alpha_0 x(t) + \alpha_1 x(t - T) + \beta u(t)$

(*d*) $\mathring{x}(t) = 0$

$y(t) = \delta u(t - T)$

These models are not, in general, derived from finite-dimensional process descriptions described in Section 2.2.1, but rather from partial differential equations. To keep matters simple our starting point will be the linearised models containing time delays. If we wish to proceed with estimator or controller design we must distinguish between situations where an approximation must be made concerning the effect of the time delay and the situation where no approximation is necessary. Into this latter category fall the following three types of time-delay model:

The *input time-delay model* is

$$\mathring{x}(t) = A(t)x(t) + B_0(t)u(t) + B_1(t)u(t - T) \tag{2.18}$$

$$x(t_0) = x_0 \tag{2.19}$$

$$y(t) = C(t)x(t) \tag{2.20}$$

The *output time-delay model* is

$$\mathring{x}(t) = A(t)x(t) + B(t)u(t) \tag{2.21}$$

$$x(t_0) = x_0 \tag{2.22}$$

$$y(t) = C_0(t)x(t) + C_1(t)x(t - T) \tag{2.23}$$

The *memoryless time-delay model* is

$$\mathring{x}(t) = A(t)x(t) + B(t)u(t) \tag{2.24}$$

$$x(t_0) = x_0 \tag{2.25}$$

$$y(t) = C(t)x(t) + D(t)u(t - T) \tag{2.26}$$

When the time delay is in one or more stable variables in the state equation (often the case in process control dynamics) the situation is more difficult and an approximation must be introduced.

The *state time-delay model* is

$$\mathring{x}(t) = A_0(t)x(t) + A_1(t)x(t - T) + B(t)u(t) \tag{2.27}$$

$$x(t_0) = x_0 \tag{2.28}$$

$$y(t) = C(t)x(t) \tag{2.29}$$

There are at least three ways of approximating the dynamic behaviour of this model so that estimator or controller design may proceed. Two of these methods, described first below, are restricted to the time-invariant case.

For simplicity consider a scalar state model subject to time delay:

$$\mathring{x}(t) = \alpha_0 x(t) + \alpha_1 x(t - T) + \beta u(t) \tag{2.30}$$

Taking Laplace transforms of both sides gives

$$(s - \alpha_0 - \alpha_1 \, \mathrm{e}^{-Ts})x(s) = \beta u(s) \tag{2.31}$$

where s is the Laplace parameter. Two approximations to e^{-Ts} can now be introduced. Either,

$$\mathrm{e}^{-Ts} = \lim_{n\to\infty} \left(1 + \frac{T}{n} s\right)^{-n} \tag{2.32a}$$

$$\simeq \left(1 + \frac{T}{n} s\right)^{-n}, \quad n \text{ large} \tag{2.32b}$$

$$\simeq \left(1 + \frac{T}{n} s\right)^{-1}\left(1 + \frac{T}{n} s\right)^{-1} \ldots \left(1 + \frac{T}{n} s\right)^{-1} \tag{2.32c}$$

or

$$\mathrm{e}^{-Ts} = \mathrm{e}^{-\alpha Ts} \, . \, \mathrm{e}^{-(1-\alpha)Ts} \tag{2.33a}$$

$$= \frac{\mathrm{e}^{-\alpha Ts}}{\mathrm{e}^{(1-\alpha)Ts}} \tag{2.33b}$$

$$\approx \frac{1 - \alpha Ts + \alpha^2 T^2 s^2/2}{1 + (1 - \alpha)Ts + (1 - \alpha)^2 T^2 s^2/2} \tag{2.33c}$$

$$\approx \frac{1 - \alpha Ts}{1 + (1 - \alpha)Ts + (1 - \alpha)^2 T^2 s^2/2}, \quad \alpha \text{ small} \tag{2.33d}$$

The first approximation uses n first-order models (see Section 2.2.5) in cascade to represent the time delay, while the second (known as the Pade approximation) uses a second-order model which has an inverse response (see Section 2.2.5). With either approximation, the dynamic equation can be transformed back to the time domain and a finite-dimensional linear state-space model, eqn. 2.14, found which represents approximately the dynamic behaviour.

Just how large n should be, or how small α, in the approximation is a question of judgment. To aid the selection, Fig. 2.3 sets out the time response of the model $y(t) = u(t - T)$ to a step change in input and the responses of the various approximations which may be made.

The third approximation to the state time-delay model is obtained by repeatedly considering a time shift of T, the time delay

$$\mathring{x}(t) = A_0(t)x(t) + A_1(t)x(t - T) + B(t)u(t) \tag{2.27a}$$

$$\mathring{x}(t - T) = A_0(t - T)x(t - T) + A_1(t - T)x(t - 2T) + B(t - T)u(t - T) \quad (2.27b)$$

$$\mathring{x}(t - 2T) = A_0(t - 2T)x(t - 2T) + A_1(t - 2T)x(t - 3T) + B(t - 2T)u(t - 2T) \quad \text{etc.} \quad (2.27c)$$

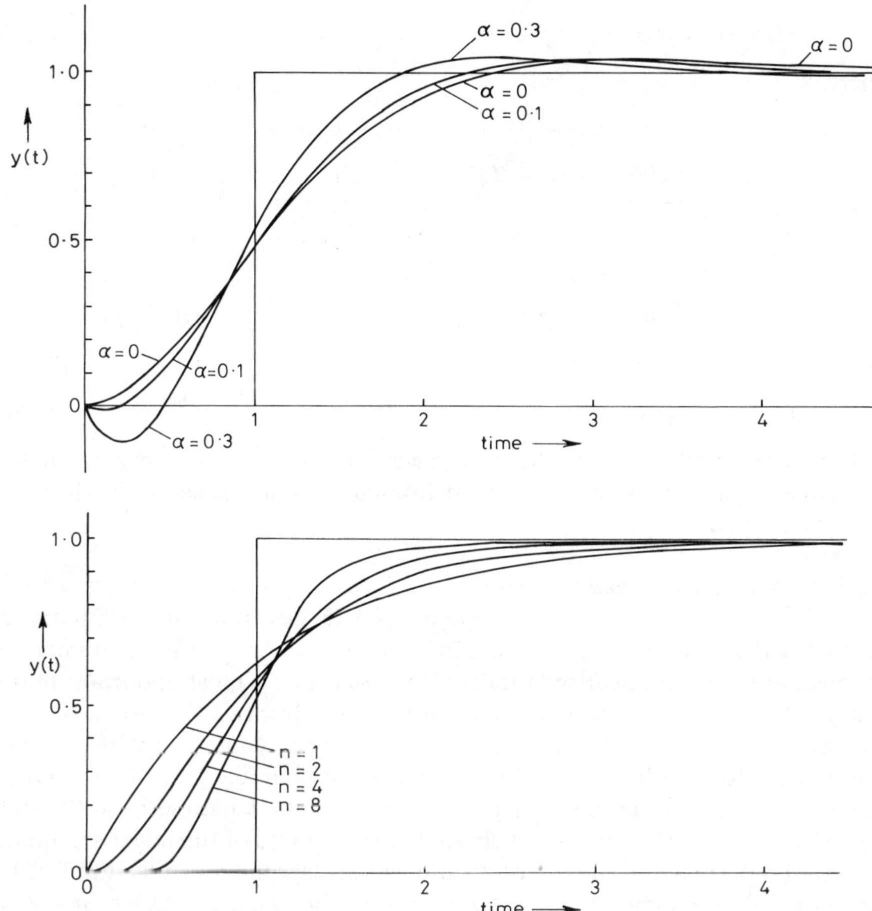

Fig. 2.3 *a* The selection of α, Pade approximation
b The selection of *n*, cascaded first-order models

until jT (where j is some integer) spans the finite time interval of interest. Assuming that $x(t) = 0$ for $t < t_0$, the last equation to be written down will be

$$\mathring{x}(t - jT) = A_0(t - jT)x(t - jT) + B(t - jT)u(t - jT) \quad (2.27d)$$

By creating an augmented state vector $z(t)$ containing $x(t)$, $x(t - T), \ldots, x(t - jT)$ we can represent the behaviour of the state time-delay model over some finite time inveral $t_0 \leqslant t \leqslant jT$ by

$$\mathring{z}(t) = \tilde{A}(t)z(t) + \text{diag}\,[B(t - iT)u(t - iT)], \quad i = 1, 2, \ldots, j \tag{2.34}$$

$$z(t_0) = z_0 \tag{2.35}$$

$$y(t) = \tilde{C}(t)z(t) \tag{2.36}$$

where

$$\tilde{A}(t) \triangleq \begin{bmatrix} A_0(t) & A_1(t) & 0 & \cdots & 0 \\ 0 & A_0(t - T) & A_1(t - T) & \cdots & 0 \\ \vdots & \vdots & \vdots & & \vdots \\ 0 & 0 & 0 & \cdots & A_0(t - jT) \end{bmatrix} \tag{2.37a}$$

and

$$\tilde{C}(t) = [C(t) \quad 0 \quad 0 \quad \cdots \quad 0] \tag{2.37b}$$

Although this model is not in the state-space form of eqn. 2.14 it is nevertheless amenable to our estimator and controller design techniques, as will be shown in the next chapter.

2.2.5 *Model reduction and responses*

A small but adequate model is a great asset in designing any estimator or controller that is to do its job well. Model reduction – reducing the number of state variables in the linearised model of the process – is most important in this context. Our approach to model reduction is straightforward and can be summarised as follows. The response of any linear model consists of two components, one to the presence of a forcing function (e.g. a control signal or a disturbance), the other due to a nonzero initial state. Holding the initial state zero ($x(t) \equiv 0$ for $t \leqslant 0$) we first simulate the response of the linearised model (or a part of it) to a unit step function applied through the control variable. This response is then compared to the step response of a model of smaller dimension (henceforth called a simple model). Least-squares curve fitting can be used to substantiate the choice of simple model. When good agreement has been obtained the responses of the two models to a nonzero initial state are compared (with no forcing function present). If there is again good agreement the simple model response is finally checked against that of the nonlinear mathematical model which describes the process under investigation.

This approach to model reduction assumes a considerable understanding of dynamics on the part of the process control engineer. It is important to note how initially the response of the linearised process model and not that of the nonlinear model is used. This aids the selection of a suitable simple model. To help matters further we now list some of the simple models for reference purposes (see Table 2.1). These model responses, to both nonzero initial state and unit

Table 2.1

	Model	Synonyms	Model equation(s)
1	Pure gain		$y(t) = \delta u(t)$
2	Pure integrator	Capacity	$\mathring{x}(t) = u(t)$
3	First order	Stirred tank	$\mathring{x}(t) = \alpha x(t) + \beta u(t)$
4	Second order (no zero)	Second order	$\mathring{x}_1(t) = x_2(t)$ $\mathring{x}_2(t) = \alpha_1 x_1(t) + \alpha_2 x_2(t) + u(t)$ $y(t) = x_1(t)$
5	General second order	(Non)minimum phase. Inverse response. Undershoot.	$\mathring{x}_1(t) = x_2(t)$ $\mathring{x}_2(t) = \alpha_1 x_1(t) + \alpha_2 x_2(t) + u(t)$ $y(t) = \gamma_1 x_1(t) + \gamma_2 x_2(t)$
6	Memoryless time delay	Dead time. Transport lag	$y(t) = u(t - T)$
7	Input time delay		$\mathring{x}(t) = \alpha x(t) + \beta u(t - T)$
8	Output time delay		$\mathring{x}(t) = \alpha x(t) + \beta u(t)$ $y(t) = x(t - T)$
9	State time delay	Retarded state	$\mathring{x}(t) = \alpha_0 x(t) + \alpha_1 x(t - T) + \beta u(t)$

step forcing function, are shown in Fig. 2.4. Table 2.2 lists the mathematical expressions for the responses, though these are usually simulated, not calculated (see Chapter 5). Brief remarks concerning some of the simple models now follow.

First-order model: If $\alpha < 0$ the model is stable (check with Table 2.2). If

$$\lim_{t\to\infty} u(t) = u,$$

a constant, then the unique steady state that is eventually attained is given by

$$x = -\frac{\beta}{\alpha} u$$

The dynamic response of the model is governed by the single, real, eigenvalue α. The response can therefore never oscillate if the input variable is nonoscillatory. With the state initially zero but with nonzero input variable, the

Table 2.2

	Model	Step Reponse*	Nonzero initial state response†
1	Pure gain	$y(t) = \delta$	—
2	Pure integrator	$x(t) = t$	$x(t) = 1.0$
3	First order		
	$\alpha \neq 0$	$x(t) = -\dfrac{\beta}{\alpha}(1 - e^{\alpha t})$	$x(t) = e^{\alpha t}$
	$\alpha = 0$	$x(t) = \beta t$	$x(t) = 1.0$
4	Second order (no zero)		
	$-4\alpha_1 < \alpha_2^2$	$y(t) = \dfrac{1}{\alpha_1}\left(e^{\frac{\alpha_2 t}{2}}\left[\cosh\left(\alpha_1 + \dfrac{\alpha_2^2}{4}\right)^{\frac{1}{2}} t - \dfrac{\alpha_2}{2\left(\alpha_1 + \dfrac{\alpha_2^2}{4}\right)^{\frac{1}{2}}} \times \sinh\left(\alpha_1 + \dfrac{\alpha_2^2}{4}\right)^{\frac{1}{2}} t\right] - 1\right)$	$y(t) = \dfrac{e^{\frac{\alpha_2 t}{2}}}{2(\alpha_2^2 + 4\alpha_1)^{\frac{1}{2}}}\left([(\alpha_2^2 + 4\alpha_1)^{\frac{1}{2}} - \alpha_2 + 2]\, e^{(\alpha_2^2 + 4\alpha_1)^{\frac{1}{2}}\frac{t}{2}} + [(\alpha_2^2 + 4\alpha_1)^{\frac{1}{2}} + \alpha_2 - 2]\, e^{-(\alpha_2^2 + 4\alpha_1)^{\frac{1}{2}}\frac{t}{2}}\right)$
	$-4\alpha_1 = \alpha_2^2$	$y(t) = \dfrac{1}{\alpha_1}[(1 + (-\alpha_1)^{\frac{1}{2}} t)\, e^{-(-\alpha_1)^{\frac{1}{2}} t} - 1]$	$y(t) = (1 + t + (-\alpha_1)^{\frac{1}{2}} t)\, e^{-(-\alpha_1)^{\frac{1}{2}} t}$
	$-4\alpha_1 > \alpha_2^2$	$y(t) = \dfrac{1}{\alpha_1}\left(e^{\frac{\alpha_2 t}{2}}\left[\cos\left(-\alpha_1 - \dfrac{\alpha_2^2}{4}\right)^{\frac{1}{2}} t\right.\right.$	$y(t) = e^{\frac{\alpha_2 t}{2}}\left[\cos\left(-\alpha_1 - \dfrac{\alpha_2^2}{4}\right)^{\frac{1}{2}} t\right.$

		$-\dfrac{\alpha_2}{2\left(-\alpha_1 - \dfrac{\alpha_2^2}{4}\right)^{\frac{1}{2}}} \times \sin\left(-\alpha_1 - \dfrac{\alpha_2^2}{4}\right)^{\frac{1}{2}} t \Bigg] - 1 \Bigg)$	$+ \dfrac{2 - \alpha_2}{(-4\alpha_1 - \alpha_2^2)^{\frac{1}{2}}} \times \sin\left(-\alpha_1 - \dfrac{\alpha_2^2}{4}\right)^{\frac{1}{2}} t \Bigg]$
5	General second order $-4\alpha_1 = \alpha_2^2$	$y(t) = \dfrac{1}{\alpha_1}[(1 + (-\alpha_1)^{\frac{1}{2}} t) \times e^{-(-\alpha_1)^{\frac{1}{2}} t} - 1] - t\, e^{-(-\alpha_1)^{\frac{1}{2}} t}$	$y(t) = e^{-(-\alpha_1)^{\frac{1}{2}} t}(2 + t[1 + \alpha_1])$
6	Memoryless time delay	$y(t) = \begin{cases} 0, & 0 \leqslant t < T \\ 1, & T \leqslant t \end{cases}$	—
7	Input time delay	$x(t) = \begin{cases} 0, & 0 \leqslant t < T \\ -\dfrac{\beta}{\alpha}(1 - e^{\alpha(t-T)}) \end{cases}$	$x(t) = e^{\alpha t}$
8	Output time delay	$y(t) = \begin{cases} 0, & 0 \leqslant t < T \\ -\dfrac{\beta}{\alpha}(1 - e^{\alpha(t-T)} \end{cases}$	$y(t) = \begin{cases} 0, & 0 \leqslant t < T \\ e^{\alpha(t-T)} \end{cases}$
9	State time delay	$x(t) = -\dfrac{\beta}{\alpha_0}(1 - e^{\alpha_0 t}), \quad 0 \leqslant t \leqslant T$	$x(t) = e^{\alpha_0 t}, \quad 0 \leqslant t \leqslant T$

* Response for $t > 0$ with $u(t) = U(t)$ and $x(0) = 0$.
† Response for $t > 0$ with $x(0) = 1$ and $u(t) \equiv 0$.

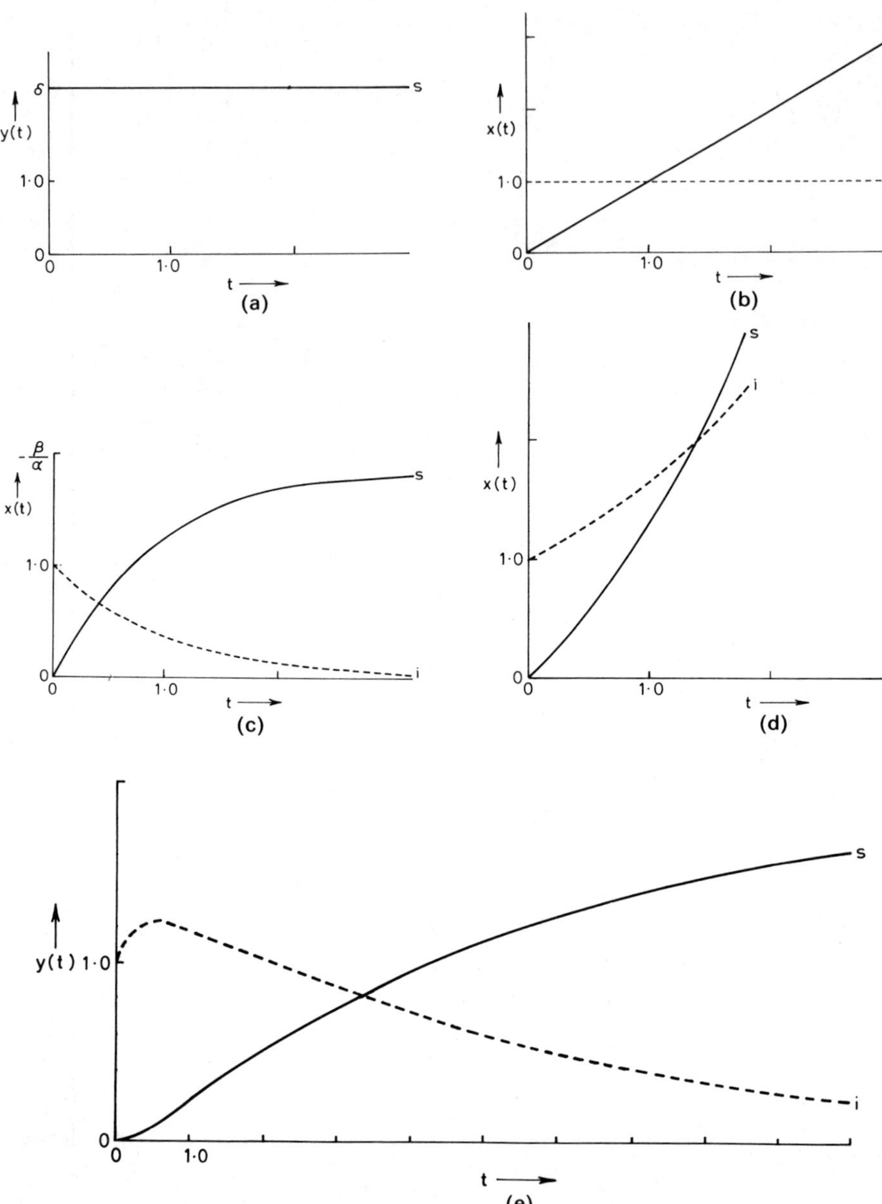

Fig. 2.4 *s* step response
i initial state response
a Pure gain
b Pure integrator
c First order, $\alpha < 0$
d First order, $\alpha > 0$
e Second order, no zero, overdamped, $-4\alpha_1 < \alpha_2^2$, $\alpha_1 < 0$ and $\alpha_2 < 0$

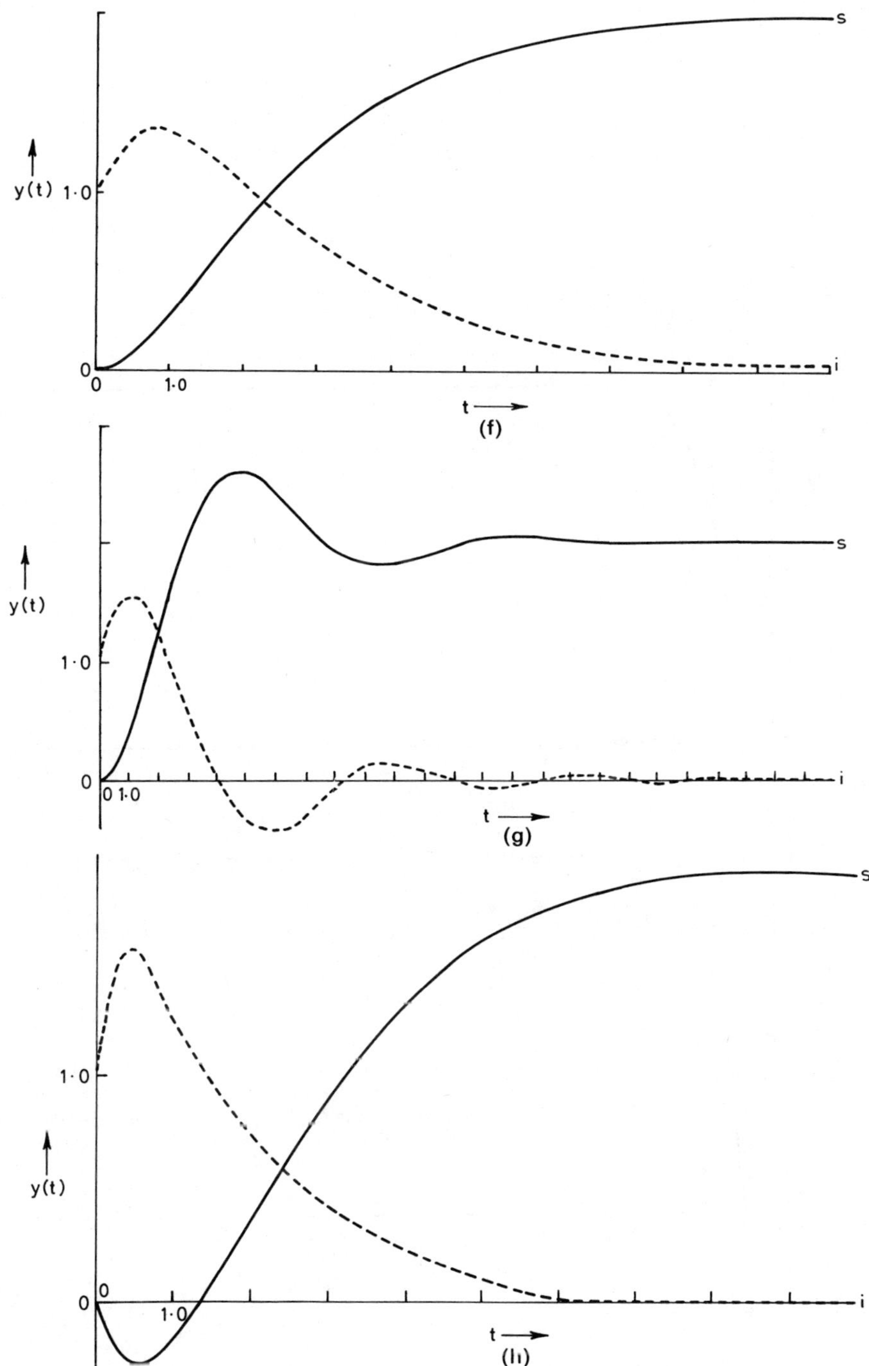

Fig. 2.4 *Continued*

f Second order, no zero, critically damped, $-4\alpha_1 = \alpha_2^2$, $\alpha_2 < 0$

g Second order, no zero, underdamped, $-4\alpha_1 > 4\alpha_2^2$, $\alpha_1 < 0$, $\alpha_2 < 0$

h General second order, inverse response, $\alpha_1 < 0$, $\alpha_2^2 = -4\alpha_1$, $\gamma_1 = 1$, $\gamma_2 = -1$

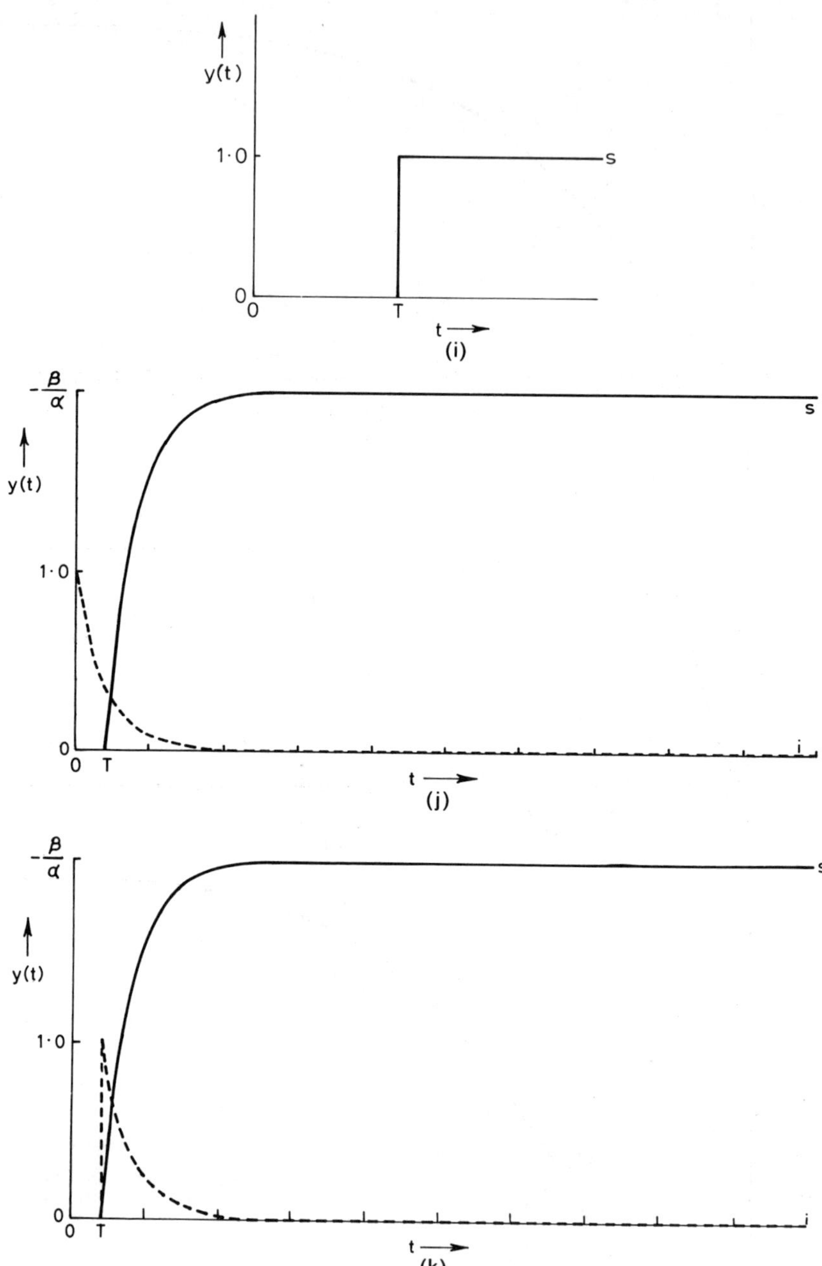

Fig. 2.4 *Continued*

i Memoryless time delay
j Input time delay, $\alpha < 0$
k Output time delay, $\alpha < 0$

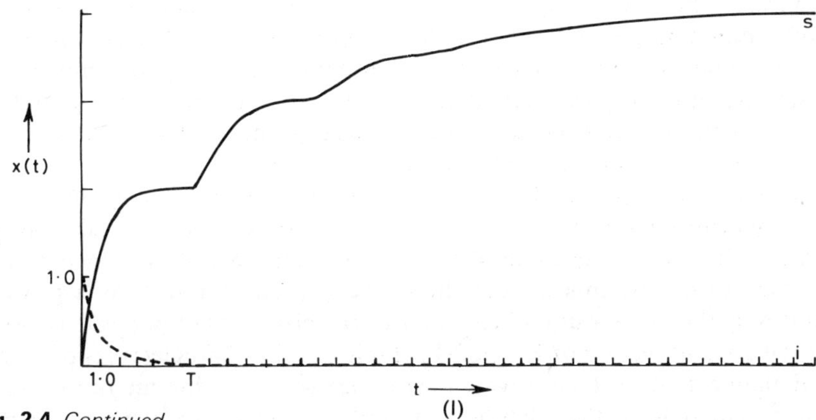

Fig. 2.4 *Continued*

l State time delay, $-\alpha_0 > \alpha_1 \geqslant 0$

initial gradient of the response is nonzero. Examples of the use of first-order models occur in mixing, fluid flow and heat transfer processes.

Second-order model (*no zero*): The model has only two free parameters α_1 and α_2. If both $\alpha_1 < 0$ and $\alpha_2 < 0$ the model is stable. In that case the unique steady state (the final value) of the model output is $-u/\alpha_1$. The model has no zero (see Chapter 1) so that the dynamic response is determined by the nature of the model's two eigenvalues.

If $-4\alpha_1 < \alpha_2^2$, both eigenvalues are real and distinct, and the response is said to be overdamped, i.e. the output can never oscillate if the input variable is nonoscillatory. When $-4\alpha_1 = \alpha_2^2$ the two eigenvalues are real but repeated, and the response is said to be critically damped. Only when $-4\alpha_1 > \alpha_2^2$ are both eigenvalues complex, and an oscillating response can result from a non-oscillatory input. Such a response is said to be underdamped.

The second-order model without any zero has an output response whose initial gradient is zero, for zero initial conditions (i.e. $x_1(t_0) = x_2(t_0) = 0$) and any input variable. This is not necessarily true for the general second-order model.

If either $\alpha_1 \geqslant 0$ or $\alpha_2 \geqslant 0$, or both, the model is unstable. The response is then either a sine wave (if $\alpha_2 = 0$ or $\alpha_1 < 0$ then the complex eigenvalues have no real parts) or a monotonically increasing function (not necessarily pure exponential) of time. There are numerous examples of second-order models without a zero: cascaded stirred tanks and some controlled processes may be adequately modelled in this way.

General second-order model: As the name implies, this model is less restrictive than the foregoing model in that there is now the possibility for one real-valued zero to exist. The zero is given by $-\gamma_1/\gamma_2$. If the zero is negative the model is said to be minimum phase. If it is non-negative the model is called nonminimum phase. Together with the eigenvalues, the zero completely determines the dynamic response. In particular, if the model is nonminimum phase and no eigenvalue is the same as the zero, the response of the model is said to be an inverse response (see Fig. 2.4). Of course, if zero and eigenvalue are the same then the dynamic response of the model degenerates to that of the first-order model. This cancellation of eigenvalue with zero shows that from the output response to the step change in input alone it will be impossible to state categorically which order dynamic model should be chosen to represent a process, except if only the input–output behaviour of the process is of interest. To some extent the general second-order model exhibits all the characteristics of higher order dynamic models. That is why it is not necessary to examine third, fourth and higher models in this chapter. Examples of processes with an inverse response include some distillation columns and evaporators.

State time-delay model: The solution $x(t)$ exists, is unique, and is stable if and only if (Mori, Fukuma and Kuwahara, 1981*b*) $-\alpha_0 > \alpha_1 \geqslant 0$. In that case the unique steady state (final value) of the model is given by $-\beta u/(\alpha_0 + \alpha_1)$. Recycle (see Example 2.2) can introduce time delays in one or more states.

2.3 Properties of continuous models

In Section 2.2 the finite-dimensional linear state-space model and various time-delay models were introduced. The simple models of the previous section are just special cases of these general models. Recapping, we had:

---

the linear, time-varying, continuous state-space model

$$\mathring{x}(t) = A(t)x(t) + B(t)u(t) \tag{2.38}$$

$$x(t_0) = x_0 \tag{2.39}$$

$$y(t) = C(t)x(t) \tag{2.40}$$

with

$$x(t) \in R^n, \quad u(t) \in R^r, \quad y(t) \in R^m$$

---

and, as a special case of the above:

---

the linear, time-invariant, continuous state-space model

$$\mathring{x}(t) = Ax(t) + Bu(t) \tag{2.41}$$

$$x(t_0) = x_0 \tag{2.42}$$

$$y(t) = Cx(t) \tag{2.43}$$

with

$$x(t) \in R^n, \quad u(t) \in R^r, \; y(t) \in R^m$$

Finally, we had:

---

the linear, time-varying, continuous time-delay model

$$\mathring{x}(t) = A_0(t)x(t) + A_1(t)x(t - T_1) + B_0(t)u(t) + B_1(t)u(t - T_2) \tag{2.44}$$

$$x(t_0) = x_0 \tag{2.45}$$

$$x(t) = 0, \quad t \leqslant t_0 \tag{2.46}$$

$$y(t) = C_0(t)x(t) + C_1(t)x(t - T_3) + D_0(t)u(t) + D_1(t)u(t - T_4) \tag{2.47}$$

with

$$x(t) \in R^n, \quad u(t) \in R^r, \quad y(t) \in R^m$$

Of course, the first two models are only special cases of the last model. The distinction is made because there has been a rather complete analysis of the linear state-space model in contrast to the linear time-delay case.

Properties of these linear models must be related to those of the nonlinear process of interest. Indeed, it is possible to formalise such relationships; Lyapunov's (First) Theorem for stability analysis (Willems, 1970) is an example. The approach recommended here is more heuristic than formal, namely that several linearisations around various reference trajectories or points are to be performed and the necessary investigations into the properties of the process be carried out on the (several) resulting linear models. Obviously a computer is of great assistance here.

A question which naturally arises is how sensitive the properties are to the various parameter values. To answer this question would go beyond the scope of this text: we remark, however, that two techniques, namely interval analysis (Moore, 1966) and structural analysis (Barnett and Storey, 1970; Lin, 1974; Davison, 1977) could be of use here.

2.3.1 *Equilibrium*

Most processes are operated in a continuous fashion (as opposed to batch or semi-batch) about an equilibrium point or steady state. Linearisation, it will be recalled, is often performed about this equilibrium point. Moreover, in the study of stability or inverse response the steady state plays an important role. Formally:

---

x is a *steady state* (equilibrium point) of a model if once the unforced state of the model is x it remains equal to x for all future time.

Let us now consider the unforced, time-varying state-space model. The state differential equation is

$$\mathring{x}(t) = A(t)x(t) \tag{2.48}$$

and from the foregoing definition a steady state must satisfy

$$0 = A(t)x(t) \tag{2.49}$$

for all t. Hence the first result, namely:

The *null* (or zero) solution, $x \equiv 0$, is always a steady state of the unforced time-varying state-space model, eqns. 2.38 to 2.40.

This result holds even for unstable models. The null solution is often not the only possible steady state.

The unforced, time-varying state-space eqn. 2.38 has a *unique* steady state $x \equiv 0$ if and only if, for any t_1 and t_2 ($> t_1$) the Gramian matrix G, where

$$G \triangleq \int_{t_1}^{t_2} A(t)A^{\mathrm{T}}(t)\,\mathrm{d}t$$

is nonsingular.

A corollary of the above is:

The unforced, time-invariant state-space eqn. 2.41 has a unique steady state $x \equiv 0$ if and only if A is nonsingular.

Obviously, if A is singular there will be an infinite number of steady states.

So far it has been assumed that there is no forcing. When a constant forcing term, $u(t) = u$, is allowed there is obviously only equilibrium if there is stability:

The time-invariant state-space eqn. 2.41 has a unique steady state

$$x = -A^{-1}Bu \tag{2.50}$$

if and only if A is stable and B has full rank.

A matrix A is said to be stable if the real parts of all its eigenvalues are negative, i.e. $\mathrm{Re}\{\lambda_i(A)\} < 0$. If A is stable, its inverse, A^{-1}, always exists. Note finally that if B does not have full rank then some of the control variables are redundant.

2.3.2 *Stability*

One of the purposes of designing a controller is to ensure that the controlled process is stable. Even if a process is stable, we may be interested to know the degree of stability or how sensitive the process stability is to parameter changes.

Roughly speaking, a stable steady state is one where small perturbations from the steady state always return to that steady state. For nonlinear processes with many steady states it would therefore be misleading to speak of the stability *of a process*. When using a linearised model of a process the stability of each steady state of *the process* must be investigated (not of the model, since the stability properties of all steady states of a linear model are the same).

Although there are many different types of stability we confine ourselves to the following (Willems, 1970):

The null solution of an unforced process is *asymptotically stable* if:

1 (boundedness) for any given t_0 and $\alpha > 0$ there exists a $\beta(\alpha, t_0) > 0$ such that if

$$\|x_0\| < \beta \text{ then } \|x(t)\| < \alpha$$

for all $t \geqslant t_0$.

2 (convergence) for any given t_0 and $\gamma > 0$ there exists a $\delta(\gamma, x_0, t_0) > 0$ and an $\varepsilon(t_0) > 0$ such that if

$$\|x_0\| < \varepsilon \text{ then } \|x(t)\| < \gamma$$

for all $t > t_0 + \delta$.

This definition appears rather complicated but reference to Fig. 2.5 should clarify matters. Notice how the norm (Chapter 1) is a rather natural choice to use here.

It turns out that the estimators and controllers we will design possess a slightly stronger form of stability than that just given. This is obtained by requiring that $\beta(\alpha, t_0)$, $\delta(\gamma, x_0, t_0)$ and $\varepsilon(t_0)$ are uniformly bounded (see Chapter 1).

The null solution of an unforced process is *uniformly asymptotically stable* if it is asymptotically stable and neither $\beta(\alpha, t_0)$, $\delta(\gamma, x_0, t_0)$ or $\varepsilon(t_0)$ depends on t_0.

Note that for time-invariant systems, uniform asymptotic stability is equivalent to asymptotic stability.

When the stability of the null solution is independent of the size of the initial deviation, x_0, we say that the process is asymptotically stable in the large, or *globally asymptotically stable*. Since all linear processes are globally asymptotically stable it is only necessary, for linear processes, to investigate the stability of the null solution. Notice also that in our definitions the process is unforced: this is because it is immaterial to the analysis just how the state became removed from the origin. We emphasise that stability is concerned with the state, not output, trajectories. Finally, before tests for stability are presented, note that from the definitions stability is never at issue in finite time problems. In the time-varying case a sufficient (but not necessary) condition for stability is the following (Mori, Fukuma and Kuwahara, 1981*a*).

The unforced, time-varying state-space model, eqns. 2.38 to 2.40, is uniformly asymptotically stable if

$$\mu[A(t)] < 0 \tag{2.51}$$

for all t.

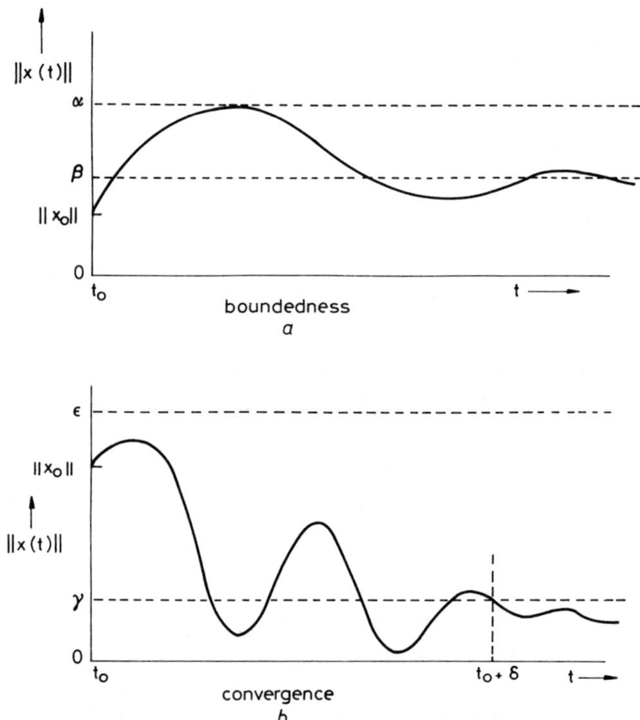

Fig. 2.5 *Asymptotic stability*

Here, $\mu[\ ]$ is the matrix measure (Chapter 1). Many other tests exist, such as those based on Lyapunov's second method, Popov's criterion etc. (Willems, 1970). However, the above test does relate to the criterion shortly to be given for the stability of time-delay models.

For time-invariant linear models a powerful yet simple test is available (Willems, 1970).

The unforced time-invariant state-space models, eqns. 2.41 to 2.43, is uniformly asymptotically stable if and only if all the eigenvalues of A have negative real parts.

Since it is only the matrix A which determines whether a time-invariant linear model is asymptotically stable, it is common to refer to A as a *stable matrix* if $\mathrm{Re}\{\lambda_i(A)\} < 0$. The Routh–Hurwitz test (see, for example, Willems, 1970) could be used to check the stability of A without explicitly evaluating its eigenvalues. Note that stability of the time-varying state-space model cannot be evaluated from the eigenvalues of $A(t)$, even if they are all constant with negative real parts (see Example 2.3).

A less well known, but simple, test when a time delay is present in the model is the following (Mori, Fukuma and Kuwahara, 1981*b*).

The unforced, time-invariant linear time-delay model, eqns. 2.44 to 2.47 with constant parameters, is uniformly asymptotically stable if

$$\mu(A_0) < -\|A_1\| \tag{2.52}$$

Here, $\|\ \|$ is the spectral norm (Chapter 1). Since $\mathrm{Re}\{\lambda_i(A_0)\} \leqslant \mu(A_0)$, the delay free case ($A_1 = 0$) reduces to requiring that A_0 be stable. Another special case is the scalar case, when eqn. 2.52 reduces to $-\alpha_0 > \alpha_1$, the necessary and sufficient condition for asymptotic stability. Note also that A_0 must be stable for eqn. 2.52 to hold, and that the criterion given above is independent of the delay, T_1.

Example 2.3
The model $\mathring{x}(t) = A(t)x(t)$ where

$$A(t) = \begin{bmatrix} \alpha & e^{\beta t} \\ 0 & \alpha \end{bmatrix}$$

has constant eigenvalues, $\lambda_1 = \lambda_2 = \alpha$. However, it is incorrect to *assume* that this time-varying linear model is asymptotically stable if $\alpha < 0$. A correct (but possibly conservative) condition for stability can be derived as follows:

$$A(t) + A^{\mathrm{T}}(t) = \begin{bmatrix} 2\alpha & e^{\beta t} \\ e^{\beta t} & 2\alpha \end{bmatrix}$$

$$\lambda\{A(t) + A^{\mathrm{T}}(t)\} = 2\alpha \pm e^{\beta t}$$

$$\mu\{A(t)\} = \alpha + \frac{e^{\beta t}}{2}$$

Then from eqn. 2.51 the model is uniformly asymptotically stable if, for all t, $-\alpha > e^{\beta t}/2$. Clearly, then, β must be negative and $\alpha < -1/2$.

Example 2.4
Consider stability of the model

$$\begin{bmatrix} \mathring{x}_1(t) \\ \mathring{x}_2(t) \end{bmatrix} = \begin{bmatrix} -1 & 0 \\ 1 & -1 \end{bmatrix} \begin{bmatrix} x_1(t) \\ x_2(t) \end{bmatrix} + \begin{bmatrix} 0 & \beta_1 \\ 0 & \beta_2 \end{bmatrix} \begin{bmatrix} x_1(t-T) \\ x_2(t-T) \end{bmatrix}$$

Since

$$A_1^{\mathrm{T}} A_1 = \begin{bmatrix} 0 & 0 \\ 0 & \beta_1^2 + \beta_2^2 \end{bmatrix}, \; \lambda_{\max}(A_1^{\mathrm{T}} A_1) = \beta_1^2 + \beta_2^2$$

so that $\| A_1 \|^2 = \beta_1^2 + \beta_2^2$. Now

$$A_0 + A_0^{\mathrm{T}} = \begin{bmatrix} -2 & 1 \\ 1 & -2 \end{bmatrix}$$

so $\lambda(A_0 + A_0^{\mathrm{T}}) = \{-1, -3\}$. Hence $\mu(A_0) = -1/2$, and inequality 2.52 becomes

$$\tfrac{1}{2} > (\beta_1^2 + \beta_2^2)^{1/2}$$

We see that the time-delay model is asymptotically stable if the parameters β_1 and β_2 satisfy the stability condition that $\beta_1^2 + \beta_2^2 < 1/4$.

Lastly in this section, some notes concerning the sensitivity of the stability to parameter variations. We begin by recalling from Chapter 1 that the stability of a time-invariant linear system is invariant under a system similarity transformation. The result can be extended to time-varying linear systems under the Lyapunov transformation (Gantmacher, 1974). As we have seen, for time-invariant linear systems the eigenvalues, or more specifically the *spectral radius* (Chapter 1), determines stability. Algorithms for computing eigenvalue sensitivity may be found in the literature (Mitchell, Nail and McDaniel, 1979; Eslami and Russell, 1980).

Related to the stability sensitivity is the *degree of stability* of a process. Suppose the matrix measure, $\mu\{A(t)\}$, of the unforced linear state-space model

$$\mathring{x}(t) = A(t)x(t)$$
$$x(t_0) = x_0$$

is negative for all t. Then the model is uniformly asymptotically stable. Changes are now made so that $A(t)$ becomes $A^1(t)$, and the matrix measure of $A^1(t)$ becomes even more negative, i.e.

$$\mu\{A^1(t)\} = \mu\{A(t)\} - \alpha \tag{2.53a}$$

$$= \mu\{A(t) - \alpha I\} \tag{2.53b}$$

where α is some positive scalar. The last equation follows from one of the properties of eigenvalues (see Chapter 1). What can be said of the response, call it $z(t)$, of the changed model? We can easily show that if

$$z(t) = e^{-\alpha t}x(t) \tag{2.54}$$

then

$$\begin{aligned} \mathring{z}(t) &= (A(t) - \alpha I)z(t) \\ &= A^1(t)z(t) \end{aligned} \tag{2.55}$$

which describes the dynamics of the changed model. Clearly the changed model response is more damped than $x(t)$. We say that the model has a degree of stability of α; the greater the degree of stability the more negative the matrix measure becomes and the faster the unforced response to a nonzero initial state decays to zero.

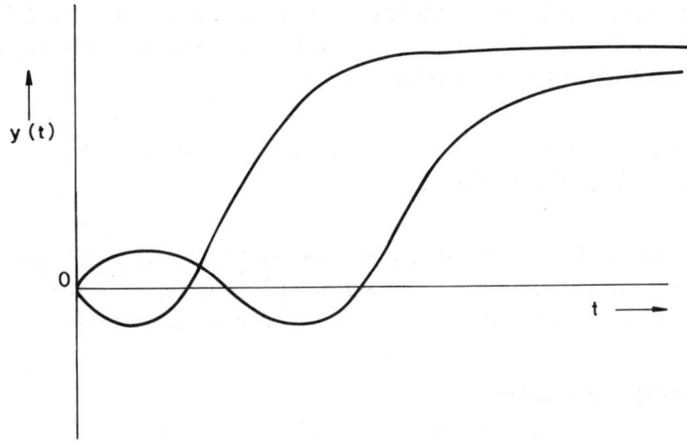

Fig. 2.6 *Types of inverse response*

2.3.3 *Inverse response*

Uncontrolled processes rarely exhibit an inverse response (or undershoot), see Fig. 2.6. When a controller is present (the process is then said to be in a closed-loop) the situation is different and inverse responses are often encountered. From a process operator's viewpoint, a process with an inverse response can be difficult to run, and the time to reach a new equilibrium (called the rise time) may be unacceptably long. Although methods exist (Mita and Yoshida, 1981; Porter, 1982) to ensure controllers do not induce an inverse response, we will not present them in this book; our design techniques for estimators and controllers allow the process control engineer sufficient freedom to influence state and output responses without the need for special measures to be taken.

Our discussion of the inverse response will be restricted (unless otherwise stated) to s.i.s.o. processes. We begin with a definition.

Let a process be subjected to a unit step function input. If, for any $t \geqslant t_0$

$$y(t) \cdot y(\infty) < 0$$

where

$$y(\infty) = \lim_{t \to \infty} y(t)$$

is the equilibrium value, then the process is said to have an *inverse response*.

Now there is a connection between an inverse response and the zeros of the process. Let us first distinguish (Davison, 1983) between two types of model, depending upon their zero positions.

The s.i.s.o. time-invariant state-space model, eqns. 2.41 to 2.43, is *non-minimum phase* if one or more of its zeros has a non-negative real part. Otherwise the model is said to be *minimum phase*.

The connection referred to above is true only for the single-input/single-output case. It is well known that:

If and only if the s.i.s.o. time-invariant state-space model, eqns. 2.41 to 2.43, is nonminimum phase and not all the zeros with non-negative real parts are the same as the model eigenvalues, it will have an inverse response.

An immediate corollary is:

If the s.i.s.o. time-invariant state-space model, eqns. 2.41 to 2.43, is non-minimum phase and asymptotically stable, then it will have an inverse response.

The above might lead us to believe that inverse reponse in multivariable processes depends on nonminimum phase and zero/eigenvalue cancellation characteristics. However, it has recently been pointed out (Mita and Yoshida, 1981) that this is not so. For multivariable, time-invariant systems we have the following definition of a minimum phase system:

The time-invariant state-space model, eqns. 2.41 to 2.43, is a minimum phase system if

$$\operatorname{rank}\begin{bmatrix} \lambda I - A & B \\ C & 0 \end{bmatrix} = n + m, \quad \operatorname{Re}(\lambda) \geqslant 0$$

This definition is equivalent to that already given for the special s.i.s.o. case.

2.3.4 *Controllability*

A fundamentally important question is whether all the state variables of a process could be transferred from some given initial state to any desired, yet feasible, terminal state. If the answer is yes, the process is said to be controllable; if just one (or more) state variables cannot be manipulated the process is uncontrollable. Clearly, then, the property of controllability is a fairly severe criterion to use in practice since many uncontrollable processes will nevertheless have quite acceptable output behaviour, with any state variables which cannot be manipulated harmlessly decaying away as time goes on. Therefore the controllability of a process is too harsh a condition to impose for satisfactory controller design. Nevertheless, without an understanding of controllability it is impossible to appreciate the requirement of stabilisability (see the next section) which is sufficient to ensure the existence of a steady-state solution to the control problem and a stable solution to the estimator problem.

When it comes to defining precisely what is meant by controllability of a process we find that, just as with stability, there are a number of definitions. We choose just one of these, since it is applicable to controller design for time-varying processes. In the time-invariant case it degenerates to the commonly used definition of controllability. Just as with stability, our definition will require that certain quantities are uniformly bounded (Chapter 1). Although the concept of controllability is introduced in a rather mathematical context the link with control soon follows.

We begin by defining a matrix of great importance in the study of controllability (Kalman, Falb and Arbib, 1969; Brockett, 1970):

The symmetric *controllability Gramian* is defined by

$$W(t_0, t_1) \triangleq \int_{t_0}^{t_1} \Phi(t_0, s)B(s)B^{\mathrm{T}}(s)\Phi^{\mathrm{T}}(t_0, s)\,\mathrm{d}s \tag{2.56}$$

where $B(t)$ is the input matrix and $\Phi(t, \tau)$ the transition matrix corresponding to $A(t)$.

If $A(t)$ and $B(t)$ are both bounded, then $W(t_0, t_1)$ is uniformly bounded from above:

Let $\|A(t)\| \leqslant \alpha$ and $\|B(t)\| \leqslant \beta$, for all t. Then

$$W(t_0, t_1) \leqslant \gamma(\delta)I \tag{2.57}$$

where $\delta = t_1 - t_0$, with $t_1 > t_0 \geqslant 0$.

Proof:

$$\Phi(t_1, t_0) = \int_{t_0}^{t_1} \dot{\Phi}(s, t_0)\,\mathrm{d}s + \Phi(t_0, t_0)$$

$$= \int_{t_0}^{t_1} A(s)\Phi(s, t_0)\,\mathrm{d}s + I$$

$$\therefore \; \|\Phi(t_1, t_0)\| \leqslant \int_{t_0}^{t_1} \|A(s)\|\,\|\Phi(s, t_0)\|\,\mathrm{d}s + I$$

$$\leqslant \exp \int_{t_0}^{t_1} \|A(s)\|\,\mathrm{d}s$$

from Gronwall–Bellman lemma (Wiberg, 1971)

$$\therefore \; \|\Phi(t_1, t_0)\| \leqslant \exp \alpha(t_1 - t_0) \tag{2.58}$$

Now,

$$\begin{aligned} \|W(t_0, t_1)\| &= \int_{t_0}^{t_1} \|\Phi(t_0, s)B(s)B^{\mathrm{T}}(s)\Phi^{\mathrm{T}}(t_0, s)\|\,\mathrm{d}s \\ &\leqslant \int_{t_0}^{t_1} \|\Phi(t_0, s)\|^2\,\|B(s)\|^2\,\mathrm{d}s \\ &\leqslant \kappa \int_{t_0}^{t_1} \mathrm{d}s, \text{ since } A(t), B(t) \text{ bounded} \\ &\leqslant \gamma(\delta) \end{aligned}$$

$$\therefore \; \lambda_{\max}(W) \leqslant \gamma, \text{ since } W(t_0, t_1) \text{ is symmetric}$$

$$\therefore \; \lambda_{\max}(W - \gamma I) \leqslant 0$$

$$\therefore \; W \leqslant \gamma I \qquad \text{QED}$$

For *uniform complete controllability* we require that the controllability Gramian is also bounded uniformly from below.

Let $\|A(t)\| \leqslant \alpha$ and $\|B(t)\| \leqslant \beta$ for all t, and let $\delta \triangleq t_1 - t_0$ where $0 \leqslant t_0 < t_1$. Then the pair $[A(t), B(t)]$ is uniformly completely controllable if

(i) $W(t_0, t_1) > 0$ (2.59*a*)

(ii) $W(t_0, t_1) \geqslant \gamma(\delta)I$ (2.59*b*)

In this definition some new terminology has been introduced. *The pair* $[A(t), B(t)]$ refer to two matrices of the state-space model

$$\begin{aligned} \mathring{x}(t) &= A(t)x(t) + B(t)u(t) \\ x(t_0) &= x_0 \\ y(t) &= C(t)x(t) \end{aligned}$$

and is a notational shorthand that will be frequently encountered henceforth.

It is the second of the two conditions in the definition which distinguishes uniform complete controllability from complete controllability. The crucial point here is that γ may depend on $t_1 - t_0$ but not on either t_1 or t_0. Note that we could have stated the definition in terms of the requirement that the eigenvalues of the controllability Gramian be uniformly bounded from above and below.

Our definition is not too useful as a test for uniform complete controllability of bounded processes, since $\Phi(t, \tau)$, the transition matrix, must be known or computed; however, this difficulty can be alleviated (Kern, 1982).

Towards establishing a link between the foregoing and control, we have the following result:

Consider the linear, time-varying state eqn. 2.38. Let $W(t_0, t_1)$ be invertible. Take

$$\tilde{u}(t) = -B^{\mathrm{T}}(t)\Phi^{\mathrm{T}}(t_0, t_1)W^{-1}(t_0, t_1)x_0 \tag{2.60}$$

where $x_0 \in R^n$ and assume that $x(t_1) = 0$. Then

$$\text{(i)} \quad x(t_0) = x_0 \tag{2.61a}$$

$$\text{(ii)} \quad \int_{t_0}^{t_1} \|\tilde{u}(s)\|^2 \, \mathrm{d}s = x_0^{\mathrm{T}} W^{-1}(t_0, t_1)x_0 \tag{2.61b}$$

$$\text{(iii)} \quad \lambda_{\min}(W) \leqslant \frac{\|x_0\|^2}{\int_{t_0}^{t_1} \|\tilde{u}(s)\|^2 \, \mathrm{d}s} \leqslant \lambda_{\max}(W) \tag{2.61c}$$

Proof:

$$\text{(i)} \quad x(t_1) = \Phi(t_1, t_0)x(t_0) + \int_{t_0}^{t_1} \Phi(t_1, s)B(s)\tilde{u}(s)\,\mathrm{d}s \quad \text{(Chapter 1)}$$

$$\therefore -\Phi(t_1, t_0)x(t_0) = \Phi(t_1, t_0)\int_{t_0}^{t_1} \Phi(t_0, s)B(s)\tilde{u}(s)\,\mathrm{d}s$$

$$= -\Phi(t_1, t_0)W(t_0, t_1)W^{-1}(t_0, t_1)x_0$$

$$x(t_0) = x_0 \qquad \text{QED}$$

$$\text{(ii)} \quad \|\tilde{u}(t)\|^2 = \tilde{u}^{\mathrm{T}}(t)\tilde{u}(t)$$

$$= x_0^{\mathrm{T}} W^{-1}(t_0, t_1)\Phi(t_0, t)B(t)B^{\mathrm{T}}(t)\Phi^{\mathrm{T}}(t_0, t)W^{-1}(t_0, t_1)x_0$$

$$\therefore \int_{t_0}^{t_1} \|\tilde{u}(s)\|^2 \, \mathrm{d}s = x_0^{\mathrm{T}} W^{-1}(t_0, t_1)W(t_0, t_1)W^{-1}(t_0, t_1)x_0$$

$$= x_0^{\mathrm{T}} W^{-1}(t_0, t_1)x_0 \qquad \text{QED}$$

$$\text{(iii)} \quad \text{From (ii)} \quad \frac{\int_{t_0}^{t_1} \|\tilde{u}(s)\|^2 \, \mathrm{d}s}{x_0^{\mathrm{T}} x_0} = \frac{x_0^{\mathrm{T}} W^{-1}(t_0, t_1)x_0}{x_0^{\mathrm{T}} x_0}$$

The RHS is Rayleigh's quotient so that

$$\lambda_{\min}(W^{-1}) \leqslant \frac{\int_{t_0}^{t_1} \|\tilde{u}(s)\|^2 \, ds}{x_0^T x_0} \leqslant \lambda_{\max}(W^{-1})$$

$$\lambda_{\max}^{-1}(W) \leqslant \frac{\int_{t_0}^{t_1} \|\tilde{u}(s)\|^2 \, ds}{x_0^T x_0} \leqslant \lambda_{\min}^{-1}(W) \qquad \text{QED}$$

These results apply to a class of processes which include uniformly completely controllable processes. The special control chosen transfers the state of the process from any initial state at t_0 given by x_0 to the origin in some finite time $t_1 - t_0$. Although not very practical, since $\Phi(t, s)$ must be computed, and certainly not the only control to transfer the state it has an interesting interpretation (Kalman, 1963): if the integral of (ii) is the total energy input, then the given control will achieve the transfer using the minimal energy. This minimum amount of energy is given by (ii) or in a different form by (iii). Actually, a stronger result than (iii) is available (Silverman and Anderson, 1968) for uniformly completely controllable processes with bounded parameters.

Consider the linear, time-varying state eqn. 2.38 with $\|A(t)\| \leqslant \alpha$ and $\|B(t)\| \leqslant \beta$, for all t. Then the pair $[A(t), B(t)]$ is uniformly completely controllable if and only if:

(i) for every $x_0 \in R^n$ and any t_1 a control $u(\tau)$ defined on $[t_0, t_1]$ exists which transfers $x(t_0) = x_0$ to $x(t_1) = 0$.

(ii) $\|u(\tau)\| \leqslant \gamma(\delta, x_0)$ for all τ where $\delta \triangleq t_1 - t_0$.

It is the second condition, requiring that the control be uniformly bounded, which distinguishes uniform complete controllability from just complete controllability. Note that γ may depend not only on δ but also on x_0. However, the basic aim is clearly to transfer the initial state to the origin in a finite time. At first this may seem a rather restrictive definition since there are many occasions when we wish to control a process to some nonzero terminal state. In fact it is not: since

$$x(t_1) = 0 = \Phi(t_1, t_0)x_0 + \int_{t_0}^{t_1} \Phi(t_1, s)B(s)u(s)\, ds \tag{2.62}$$

$$-\Phi(t_1, t_0)x_0 = 0 + \int_{t_0}^{t_1} \Phi(t_1, s)B(s)u(s)\, ds \tag{2.63}$$

thus the control which transfers x_0 to zero also transfers the state at the origin to the new state $-\Phi(t_1, t_0)x_0$ at time t_1. Any initial state can be transferred to any final state lying within the state space.

It might be inferred from the foregoing discussion that uniform complete controllability is a rather complicated concept which is difficult to check. So as to give a fair picture we now show how relatively easy the situation becomes when time-invariant state-space models are our only concern. The first simplification is that uniform complete controllability then becomes plain controllability* as usually defined (see, for example, Chen, 1970; Kwakernaak and Sivan, 1972):

The pair (A, B) is uniformly completely controllable if and only if it is controllable.

There are many tests for controllability, so we collect them into the following result.

The pair (A, B) is controllable if any one of the following holds:

1 Rank $[B\ AB \ldots A^{n-1}B] = n$.
2 $w^{\mathrm{T}} e^{At} B = 0$ for all t implies $w = 0$.
3 $w^{\mathrm{T}} B = 0$ for some constant λ implies $w = 0$.
4 An $(n \times r)$ matrix K exists such that the eigenvalues of $A + BK^{\mathrm{T}}$ can take on arbitrary values.
5 Rank $[A - \lambda_i I\ B] = n$ where λ_i, $i = 1, 2, \ldots, n$ are the eigenvalues of A.

Proofs of these tests may be found in the literature (see, for example, Chen, 1970). For high ($n \gg 10$) order systems the determination of controllability is not a trivial problem, owing to ill conditioning of matrices, the size of matrices involved etc. Numerical considerations are very important but for details of algorithms the reader is referred elsewhere (Davison, Gesing and Wang, 1978; Paige, 1981; Patel, 1981).

Example 2.5
Consider the hydrolysis of an ester with caustic soda in an ideally mixed, tank reactor. The first-order reaction is of the form

$$\mathrm{R.COOR'} + \mathrm{NaOH} \rightarrow \mathrm{RCOONa} + \mathrm{R'OH}$$

where RCOOR′ is the ester A and R′OH the product B. If the reaction rate constant (l/mol.s) is given by K, then a mathematical model, linearised about an equilibrium point, would be

$$\frac{\mathrm{d}}{\mathrm{d}t}\begin{bmatrix} \Delta C_{\mathrm{A}} \\ \Delta C_{\mathrm{B}} \end{bmatrix} = \begin{bmatrix} \dfrac{-q_{\mathrm{ref}} - K}{V} & 0 \\ K & \dfrac{-q_{\mathrm{ref}}}{V} \end{bmatrix} \begin{bmatrix} \Delta C_{\mathrm{A}} \\ \Delta C_{\mathrm{B}} \end{bmatrix} + \begin{bmatrix} \dfrac{C_{\mathrm{Ai}} - C_{\mathrm{Aref}}}{V} \\ \dfrac{C_{\mathrm{Aref}} - C_{\mathrm{Ai}}}{V} \end{bmatrix} \Delta q \tag{2.64}$$

* Sometimes called complete controllability.

where

q(l/s) is the in/output flowrate

ΔC_A(mol/l) is the change in ester concentration

ΔC_B(mol/l) is the change in product concentration

V(l) is the volume of liquid in the tank

C_{Ai}(mol/l) is the input concentration of ester

According to the first test for controllability the rank of the matrix $[B\ AB]$ should be two. An easy computation shows that this is not so – the process is therefore uncontrollable.

Next some topics concerning controllability are mentioned. First, *controllability under control constraints* (Barmish and Schmitendorf, 1980; Herz, 1981). We have seen that uniform complete controllability requires that $u(t)$ be uniformly bounded from above. However, the bound depends on x_0, the initial state, so that there is no guarantee that a given constrained control could always transfer the state as desired. Hence we are led to investigate controllability under control constraints.

Another aspect of controllability deserving our attention is the fact that none of the tests given so far indicates how near to a controllable system an uncontrollable system is. For example, during the design stage of a process fabrication it may be necessary to make a choice between two competing input (i.e. manipulated) variables. Of course, the cost of the actuators involved plays a role but so do the consequences for the process dynamics. One *measure of controllability* which has been proposed (Paige, 1981) is to compute the minimum of $\|(\Delta A, \Delta B)\|$ such that the pair $(A + \Delta A, B + \Delta B)$ is uncontrollable. Other measures have also been proposed (see, for example, Rhodes, 1981).

An interesting and quite fundamental question concerning controllability is whether it is always possible to split any system into a controllable subsystem and an uncontrollable subsystem. The following result (Kalman, 1963) answers this question; it is quoted here only for the time-invariant case, although it applies more generally.

Given the linear time-invariant state-space model, eqns. 2.41 to 2.43, it is always possible to find a system similarity transformation (Chapter 1) such that

$$T^{-1}AT = \begin{bmatrix} A_{11} & A_{12} \\ 0 & A_{22} \end{bmatrix} \quad \text{and} \quad T^{-1}B = \begin{bmatrix} B_1 \\ 0 \end{bmatrix} \tag{2.65}$$

where A_{11} is a square matrix and the pair (A_{11}, B_1) is controllable.

The actual form of the transformation matrix T is given elsewhere (Kwakernaak and Sivan, 1972). The significance of the result lies in the decomposition of an uncontrollable system; note that it is now possible to label certain states or eigenvalues as being controllable or not.

Finally in this section we indicate the connection between controllability and reconstructibility (see Section 2.3.6). Kalman first stated the *duality* of various system properties (Kalman, 1960; Kalman, Falb and Arbib, 1969). We begin by defining the dual of a system (Chen, 1970).

The dual of the continuous linear time-varying state-space model, eqns. 2.38 to 2.40, is the system:

$$\mathring{z}(t) = -A^{-\mathrm{T}}(t)z(t) + C^{\mathrm{T}}(t)\tilde{u}(t) \tag{2.66}$$

$$z(t_0) = z_0 \tag{2.67}$$

$$\tilde{y}(t) = B^{\mathrm{T}}(t)z(t) \tag{2.68}$$

The connection between controllability and reconstructibility is then the following:

The continuous linear time-varying state-space model, eqns. 2.38 to 2.40, is uniformly completely controllable if and only if its dual, eqns. 266 to 2.68 is uniformly completely reconstructible.

Actually there is a further duality of properties between reachability and observability:

controllability – reconstructibility

reachability – observability

However, for linear continuous (but not for discrete) processes, controllability implies and is implied by reachability, while reconstructibility implies and is implied by observability. For time-invariant situations the concept of duality of properties reduces to:

The pair (A, B) is controllable if and only if the pair $(B^{\mathrm{T}}, A^{\mathrm{T}})$ is observable.

In conclusion, it would be gratifying to be able to extend the substantial body of controllability results to include time-delay models. However, although there is a considerable literature available (e.g. Bhat and Koivo, 1976; Manitius and Triggiani, 1978; Lee and Olbrot, 1981) the link between the design of controllers and time-delay controllability is not sufficiently developed to warrant inclusion of this topic here.

2.3.5 *Stabilisability*

We introduce this property of a process by giving a definition (Wonham, 1968).

The pair (A, B) is *stabilisable* if there exists an $(r \times n)$ real constant matrix K such that $(A - BK)$ is stable. K is said to be stabilising for the pair (A, B).

Clearly this definition applies only to time-invariant models or processes and it arises naturally from the study of the application of constant negative state feedback. Stabilisability is important in controller design, and is closely linked to the controllability (see the previous section) of the process. We recall that a process is controllable when each state of the process can be manipulated to any desired value independently of any other state. It is a rather harsh condition. However, the requirement of controllability can often be weakened to that of stabilisability. In a stabilisable process any controllable states must be stable, i.e. harmlessly decay away.

For continuous-time processes it is difficult to extend the concept of stabilisability to time-varying processes. Some results for time-varying discrete-time processes are known and will be presented in the next chapter (Section 3.4.4). Here we restrict our discussion to time-invariant processes and models.

The link between stabilisability and controllability may be formally stated as follows (Kwakernaak and Sivan, 1972).

With

$$A = \begin{bmatrix} A_{11} & A_{12} \\ 0 & A_{22} \end{bmatrix}, \qquad B = \begin{bmatrix} B_1 \\ 0 \end{bmatrix} \tag{2.69}$$

and the pair (A_{11}, B_1) controllable, the pair (A, B) is stabilisable if and only if A_{22} is a stable matrix.

Conditions ensuring (A_{11}, B_1) to be controllable have been given in Section 2.3.4. Most of the tests for stabilisability (Popeea and Lupas, 1977; Sima, 1981) work by splitting off the uncontrollable states of the system and checking whether these have associated eigenvalues such that $\mathrm{Re}\{\lambda\} < 0$. See also Section 3.3.4.

Two obvious corollaries of the above are:

If the pair (A, B) is controllable, then the pair (A, B) is stabilisable.

---

and

If the matrix A is stable, the pair (A, B) is stabilisable.

From our definition of stabilisability it follows that providing the time-invariant linear process is either controllable or stable it is always possible to ensure (at least) a stable closed-loop system using state feedback.

For processes or models with time delay the situation is much more complicated. We need an understanding of minimum phase systems and reconstructibility (Sections 2.3.3 and 2.3.6, respectively) to appreciate the following result (Furukawa and Shimeura, 1983) which is included here for completeness.

Given the linear time-invariant time-delay model, eqns. 2.44 to 2.47, with $B_1 = 0$, let $A_1 = LC$, where $L(n \times m)$ and $C(m \times n)$ are real constant matrices such that (C, A_0) is observable and rank $C = m$. Then if

(i) (A_0, B_0) is controllable

(ii) (A_0, B_0, C) is minimum phase

it is possible to find a matrix K such that the closed-loop system

$$\mathring{x}(t) = [A_0 + B_0K]x(t) + A_1x(t - T_1) \tag{2.70}$$

is asymptotically stable.

2.3.6 *Reconstructibility*

We have already mentioned in Section 2.3.4 that a dual of controllability existed called reconstructibility. This process property is concerned with answering the important question of whether all the state variables of the process can be reconstructed, or estimated, from the available data. If the answer is yes, the process is said to be reconstructible; if, however, just one (or more) state variables cannot be reconstructed the process is unreconstructible. Just as with controllability, reconstructibility is a rather severe criterion to use in practice, and we weaken the requirements in the next section, Section 2.3.7, to that of detectability. Detectability, it turns out, is sufficient to guarantee a stable controlled process and also the existence of a steady-state solution to the estimator problem.

By reconstructibility we mean, roughly speaking, whether it would ever be possible, given measurements $y(t)$ and control $u(t)$ over an interval of time (t_0, t_1), to reconstruct from them the value of the state $x(t_1)$ at time t_1. Of course, the value of the state $x(t_1)$ means the values of all the state variables – hence reconstructibility is sometimes referred to in the literature as complete reconstructibility. In agreement with the choice made in the previous section for the definition of controllability there is just one type of reconstructibility which need concern us; moreover, we will once again choose to introduce it in a rather abstract way (since it is a system property independent of $y(t)$, $u(t)$ etc.) and later show the connection with state estimation. We begin with the dual of a definition appearing in the previous section.

The symmetric *reconstructibility Gramian* (Kalman, Falb and Arbib, 1969) is

$$M(t_0, t_1) \triangleq \int_{t_0}^{t_1} \Phi^{\mathrm{T}}(s, t_1)C^{\mathrm{T}}(s)C(s)\Phi(s, t_1)\,\mathrm{d}s \tag{2.71}$$

Just as before, if $A(t)$ and $C(t)$ are uniformly bounded then $M(t_0, t_1)$ is uniformly bounded from above.

Let $\|A(t)\| \leqslant \alpha$ and $\|C(t)\| \leqslant \beta$, for all t. Then

$$M(t_0, t_1) \leqslant \gamma(\delta)I \tag{2.72}$$

where $\delta \triangleq t_1 - t_0$.

The proof is so similar to that given in the previous section that it is omitted. We next give our definition of *uniform complete reconstructibility*.

Let $\|A(t)\| \leqslant \alpha$ and $\|C(t)\| \leqslant \beta$ for all t, and let $\delta \triangleq t_1 - t_0$ where $0 \leqslant t_0 < t_1$. Then the pair $[C(t), A(t)]$ is uniformly completely reconstructible if

(i) $M(t_0, t_1) > 0$ (2.73*a*)

(ii) $M(t_0, t_1) \geqslant \gamma(\delta)I$ (2.73*b*)

Note that it is the second condition which distinguishes uniform complete reconstructibility from complete reconstructibility. To establish the link between the reconstructibility Gramian and state estimation we need the following results:

Consider the linear time-varying state-space model, eqns. 2.38 to 2.40, with $u(t) \equiv 0$. Let $M(t_0, t_1)$ be invertible. Then

(i) $$x(t_1) = M^{-1}(t_0, t_1)\int_{t_0}^{t_1} \Phi^T(s, t_1)C^T(s)y(s)\,ds \tag{2.74a}$$

(ii) $$\int_{t_0}^{t_1} \|y(s)\|^2\,ds = x^T(t_1)M(t_0, t_1)x(t_1) \tag{2.74b}$$

(iii) $$\lambda_{min}(M) \leqslant \frac{\int_{t_0}^{t_1} \|y(s)\|^2\,ds}{\|x(t_1)\|^2} \leqslant \lambda_{max}(M) \tag{2.74c}$$

Proof:

(i) $y(t) = C(t)\Phi(t, t_1)x(t_1)$ Chapter 1

$$\therefore \Phi^T(t, t_1)C^T(t)y(t) = \Phi^T(t, t_1)C^T(t)C(t)\Phi(t, t_1)x(t_1)$$

$$\therefore \int_{t_0}^{t_1} \Phi^T(s, t_1)C^T(s)y(s)\,ds = M(t_0, t_1)x(t_1) \qquad \text{QED}$$

(ii) $$\begin{aligned} \|y(t)\|^2 &= y^T(t)y(t) \\ &= x_0^T\Phi^T(t, t_0)C^T(t)C(t)\Phi(t, t_0)x_0 \\ &= x_0^T\Phi^T(t_1, t_0)\Phi^T(t, t_1)C^T(t)C(t)\Phi(t, t_1)\Phi(t_1, t_0)x_0 \\ &= x^T(t_1)\Phi^T(t, t_1)C^T(t)C(t)\Phi(t, t_1)x(t_1) \qquad \text{QED} \end{aligned}$$

(iii) From (ii) $$\frac{\int_{t_0}^{t_1} \| y(s) \|^2 \, ds}{x^T(t_1)x(t_1)} = \frac{x^T(t_1)M(t_0, t_1)x(t_1)}{x^T(t_1)x(t_1)}$$

$$\therefore \lambda_{\min}(M) \leqslant \frac{\int_{t_0}^{t_1} \| y(s) \|^2 \, ds}{x^T(t_1)x(t_1)} \leqslant \lambda_{\max}(M) \qquad \text{QED}$$

using Rayleigh's quotient formula.

These results apply to a class of processes which include uniformly completely reconstructible processes. The condition that $u(t)$ be zero on $t \in (t_0, t_1)$ may at first seem rather restrictive; in fact it is not as long as $u(t)$ is a known function of time. Just define a new measurable variable $\tilde{y}(t)$ by

$$\tilde{y}(t) \triangleq y(t) - C(t) \int_{t_0}^{t_1} \Phi(t, s)B(s)u(s) \, ds \tag{2.75}$$

and proceed as if $u(t)$ were identically zero. Just as

$$\int_{t_0}^{t_1} \| u(s) \|^2 \, ds$$

represents the total energy input in the study of controllability, so here

$$\int_{t_0}^{t_1} \| y(s) \|^2 \, ds$$

represents the total information content of the measurement. Notice also that using (i) a (rather impractical) state estimator could be constructed. It is now possible to give an equivalent but more practically oriented definition of uniform complete reconstructibility for processes with uniformly bounded parameters.

Consider the linear time-varying state-space model, eqns. 2.38 to 2.40, with $\delta \triangleq t_1 - t_0$ and with $\| A(t) \| \leqslant \alpha$ and $\| C(t) \| \leqslant \beta$ for all t. Then the pair $[C(t), A(t)]$ is uniformly completely reconstructible if and only if

(i) for every $x(t_1) \in R^n$ and any t_1 it is possible to compute $x(t_1)$ from a knowledge of $y(\tau)$ and $u(\tau)$ on $[t_0, t_1]$.

(ii) $$\frac{\int_{t_0}^{t_1} \| y(s) \|^2 \, ds}{\| x(t_1) \|^2} \geqslant \eta(\delta)$$

Proof:

(i) follows from the previous result. (ii) corresponds to the condition that $M(t_0, t_1) \geqslant \gamma(\delta)I$ since

$$M(t_0, t_1) \geqslant \gamma(\delta)I$$

$$\therefore \lambda_i(M - \gamma(\delta)I) \geqslant 0 \quad \text{for all } i$$

$$\therefore \lambda_i(M) - \gamma(\delta) \geqslant 0$$

$$\lambda_i(M) \geqslant \gamma(\delta)$$

but from the previous result

$$\frac{\int_{t_0}^{t_1} \| y(s)\|^2 \, \mathrm{d}s}{\| x(t_1)\|^2} \geqslant \lambda_{\min}(M) \geqslant \gamma(\delta)$$

Proof of necessity follows that proof given in (Silverman and Anderson, 1968) for uniform complete controllability and is not repeated here. QED

This definition has a less obvious interpretation than was the case with the dual definition for uniform complete controllability and will not be pursued further here.

We now turn to time-invariant state-space models. As with controllability, the situation simplifies considerably.

The pair (C, A) is uniformly completely reconstructible if and only if it is reconstructible*.

Recall also, from Section 2.3.4, that reconstructibility implies and is implied by observability – as long as we restrict our attention to *continuous* linear processes. Hence in the literature an 'observable pair (C, A)' may be encounted when strictly a 'reconstructible pair (C, A)' is correct. The following tests are available for reconstructibility:

The pair (C, A) is reconstructible if any one of the following holds:

1 Rank $[C^{\mathrm{T}} \; A^{\mathrm{T}}C^{\mathrm{T}} \; \ldots \; (A^{\mathrm{T}})^{n-1}C^{\mathrm{T}}] = n$.

2 $w^{\mathrm{T}}\mathrm{e}^{A^{\mathrm{T}}t}C^{\mathrm{T}} = 0$ for all t implies $w = 0$.
3 $w^{\mathrm{T}}C^{\mathrm{T}} = 0$ and $w^{\mathrm{T}}A^{\mathrm{T}} = \lambda w^{\mathrm{T}}$ for some constant λ implies $w = 0$.
4 A $(m \times n)$ matrix K exists such that the eigenvalues of $A + K^{\mathrm{T}}C$ can take on arbitrary values.
5 Rank $[A^{\mathrm{T}} - \lambda_i I \; C^{\mathrm{T}}] = n$ where $\lambda_i = 1, 2, \ldots, n$ are the eigenvalues of A.

Many of the remarks concerning controllability apply also to reconstructibility because of the dual nature of these properties. The following results are analogous to those already given in Section 2.3.4.

* Sometimes called completely reconstructible.

Given the linear time-invariant state-space model, eqns. 2.41 to 2.43, then it is always possible to find a system similarity transformation (Chapter 1) such that

$$T^{-1}AT = \begin{bmatrix} A_{11} & 0 \\ A_{21} & A_{22} \end{bmatrix} \qquad CT = [C_1 \quad 0]$$

where A_{11} is a square matrix and the pair (C_1, A_{11}) is reconstructible.

It is convenient to remark here that in fact any linear time-varying state-space model may be transformed by means of a system similarity transformation into a canonical form where both the controllable and reconstructible states are displayed (Kalman, 1963)

2.3.7 *Detectability*

Just as stabilisability is an important property of a process for control purposes, detectability of a process is vitally important in estimator design. For time-invariant processes (Wonham, 1968):

The pair (C, A) is *detectable* if there exists a $(n \times m)$ real constant matrix K such that $(A - KC)$ is stable.

The same difficulties as with stabilisability arise when we try to generalise this definition to the time-varying case; the results that are available (Anderson and Moore, 1981) concern discrete-time systems and are presented in Section 3.4.4.

There is a duality (Section 2.3.4) between detectability and stabilisability:

The pair (B, A) is detectable if the pair (A^T, B^T) is stabilisable.

Hence all the remarks given in Section 2.3.5 apply, with the appropriate modification, here. For example, we expect a close link between detectability and reconstructibility; in fact detectability gives us the assurance that any unreconstructible states are stable, or formally (Kwakernaak and Sivan, 1972):

---

with

$$A = \begin{bmatrix} A_{11} & 0 \\ A_{21} & A_{22} \end{bmatrix}, \qquad C = [C_1 \quad 0]$$

and (C_1, A_{11}) reconstructible, the pair (C, A) is detectable if and only if A_{22} is a stable matrix.

It follows that if a process is reconstructible, or if it is stable, it is also detectable.

References

ANDERSON, B. D. O. and MOORE, J. B. (1981): 'Dectability and stabilizability of time-varying discrete-time linear systems', *SIAM J. Control & Opt.*, **19**, 1, pp. 20–32

BARMISH, B. R. and SCHMITENDORF, W. E. (1980): 'A necessary and sufficient condition for local constrained controllability of a linear system', *IEEE Trans. Auto. Control*, **AC-25**, 1, pp. 97–100

BARNETT, S. and STOREY, C. (1970): 'Matrix methods in stability theory' (Thomas Nelson and Sons Ltd.)

BHAT, K. P. M. and KOIVO, H. N. (1976): 'Modal characterization of controllability and observability in time delay systems', *IEEE Trans. Auto. Control*, **AC-21**, 2, pp. 292–293

BROCKETT, R. W. (1970): 'Finite dimensional linear systems' (John Wiley)

CHEN, C. T. (1970): 'Introduction to linear system theory' (Holt, Reinhart and Winston)

DAVISON, E. J. (1977): 'Connectability and structural controllability of composite systems', *Automatica*, **13**, pp. 109–123

DAVISON, E. J. (1983): 'Some properties of minimum-phase systems and "squared-down" systems', *IEEE Trans. Auto. Control*, **AC-28**, 2, pp. 221–222

DAVISON, E. J., GESING, W. and WANG, S. H. (1978): 'An algorithm for obtaining the minimal realization of a linear time-invariant system and determining if a system is stabilizable-detectable', *IEEE Trans. Auto Control*, **AC-023**, 6, pp. 1048–1054

DAVISON, E. J. and WANG, S. H. (1975): 'New results on the controllability and observability of general composite systems', *IEEE Trans. Auto. Control*, **AC-20**, 1, pp. 123–128

DESOER, C. A. and VIDYASAGAR, M. (1975): 'Feedback systems: input–output properties' (Academic Press)

DOUGLAS, J. M. (1972): 'Process dynamics and control: Vol. 1: Analysis of dynamic systems' (Prentice-Hall)

ESLAMI, M. and RUSSELL, D. L. (1980): 'On stability with large parameter variations', *IEEE Trans. Auto. Control*, **AC-25**, 6, pp. 1231–1234

FASOL, K. H. and JORGL, H. P. (1980): 'Principles of model building and identification', *Automatica*, **16**, pp. 505–518

FRIEDLY, J. C. (1972): 'Dynamic behaviour of processes' (Prentice-Hall)

FURUKAWA, T. and SHIMEMURA, E. (1983): 'Stabilizability conditions by memoryless feedback for linear systems with time-delay', *Int. J. Control*, **37**, 3, pp. 533–565

GANTMACHER, F. R. (1974): 'The theory of matrices' (Chelsea Publishing Co., New York)

HIMMELBLAU, D. M. and BISCHOF, K. B. (1968): 'Process analysis and simulation: deterministic systems' (Wiley)

HERZ, B. (1981): 'Spectral characterization of controllability of linear time invariant systems under control restraints', *in* 'Advances in control systems and signal processing', Vol. 2, F. VIEWEG and SOHN (Braunschweig)

KALMAN, R. E. (1960): 'On general theory of control systems', Proc. 1st IFAC Int. Congress. Moscow, 1960, pp. 481–492

KALMAN, R. E., FALB, P. L. and ARBIB, M. A. (1969): 'Topics in mathematical system theory' (McGraw-Hill)

KALMAN, R. E. (1963): 'Mathematical description of linear dynamical systems', *SIAM J. Control*, **1**, 2, pp. 152–192

KERN, G. (1982): 'Uniform controllability of a class of linear time-varying systems', *IEEE Trans. Auto. Control*, **AC-27**, 1, pp. 208–210

KWAKERNAAK, H. and SIVAN, R. (1972): 'Linear optimal control systems' (John Wiley)

LEE, E. B. and OLBROT, A. (1981): 'Observability and related structural results for linear hereditary systems', *Int. J. Control*, **34**, 6, pp. 1061–1078

LIN, C. T. (1974): 'Structural controllability', *IEEE Trans. Auto. Control*, **AC-19**, 3, pp. 201–208

MANITIUS, A. and TRIGGIANI, R. (1978): 'Sufficient conditions for function space controllability and feedback stabilizability of linear retarded systems', *IEEE Trans. Auto. Control*, **AC-23**, 4, pp. 659–665

MITA, T. and YOSHIDA, H. (1981): 'Undershooting phenomenon and its control in linear multivariable servomechanisms', *IEEE Trans. Auto. Control*, **AC-26**, 2, pp. 402–407

MITCHELL, J. R., NAIL, J. B. and MCDANIEL, W. L. (1979): 'Computation of cofactors of [sI-A] with applications', *IEEE Trans. Auto. Control*, **AC-24**, 4, pp. 637–638

MOORE, R. E. (1966): 'Interval analysis' (Prentice-Hall)

MORI, T., FUKUMA, N. and KUWAHARA, M. (1981*a*): 'A stability criterion for linear time-varying systems', *Int. J. Control*, **34**, 3, pp. 585–591

MORI, T., FUKUMA, N. and KUWAHARA, M. (1981*b*): 'Simple stability criterion for single and composite linear systems with time delays, *Int. J. Control*, **34**, 6, pp. 1175–1184

PAIGE, C. C. (1981): 'Properties of numerical algorithms related to computing controllability', *IEEE Trans. Auto. Control*, **AC-26**, 1, pp. 130–138

PATEL, R. (1981): 'Computation of matrix fraction descriptions of linear time-invariant systems', *IEEE Trans. Auto. Control*, **AC-26**, 1, pp. 148–161

POPEEA, C. and LUPAS, L. (1977): 'Numerical stabilizability tests by a matrix sign function', *IEEE Trans. Auto. Control*, **AC-22**, 4, pp. 654–656

PORTER, B. (1982): 'Design of non-undershooting linear multivariable high-gain servomechanisms', *Int. J. Control*, **35**, 1, pp. 189–191

RHODES, I. B. (1981): 'Some quantitative measures of controllability and observability and their implications', Proc. 8th IFAC Int. Congress, Kyoto, Japan, pp. 27–32

ROFFEL, B. and RIJNSDORP, J. E. (1982): 'Process dynamics, control and protection' (Ann Arbor Science)

SILVERMAN, L. M. and ANDERSON, B. D. O. (1968): 'Controllability, observability and stability of linear systems', *SIAM J. Control*, **6**, 1, pp. 121–130

SIMA, V. (1981): 'An efficient Schur method to solve the stabilizing problem', *IEEE Trans. Auto. Control*, **AC-26**, 3, pp. 724–725

WIBERG, D. M. (1971): 'State space and linear systems' (McGraw-Hill)

WILLEMS, J. L. (1970): 'Stability theory of dynamical systems' (Thomas Nelson)

WONHAM, W. M. (1968): 'On a matrix Riccati equation of stochastic control', *SIAM J. Control*, **6**, 4, pp. 681–697

Chapter 3

Discrete-time process models

3.1 Introduction

A controller or an estimator is nowadays usually implemented on a digital computer of one form or another. The computer works with discrete (with respect to time) variables and it is logical, as well as easier, to develop control and estimation algorithms in the discrete-time domain. Before the continuous-time sensor output (discrete-time measurements are encountered less commonly) can be accepted by the computer it must be sampled. Moreover, the sequence of numbers which is the computer output must be suitably processed (*signal reconstruction*) before being sent to the actuators of the process since these are usually analogue (i.e. continuous-time) devices.

The purpose of this chapter is to analyse the dynamics of the sampling and signal reconstruction operations just described and to show how it is possible to create a comprehensive discrete-time model which describes not only these operations but also the behaviour of the continuous-time process at the sampling instants. This model – which we will call the equivalent discrete system or EDS for short – will be the basis for the controller and estimator design techniques of later chapters. In this way the controller or estimator will not only take into account the process dynamics but also the sampling and reconstruction effects.

Just as with the continuous-time models of the previous chapter, we must allow the EDS to be stochastic if a worthwhile and realistic model is to be obtained. We show how this can be achieved using additive discrete-time white noise processes.

On the practical side the topics of high-frequency noise suppression and the choice of sampling period are discussed. Before a controller or estimator can be designed both of these questions must be resolved.

In Section 3.3 a number of responses of simple EDSs are given. These discrete systems correspond to the sampled forms of the continuous-time process models whose step responses were investigated in Section 2.2.5. A comparison of step responses shows that at the sampling instants they are the same; however, the behaviour of the EDS between the sampling instants is not defined.

Lastly in this chapter a number of properties of the EDS in its most general form are considered. Some of these properties such as the mean and covariance are essential for a statistical description of the behaviour of the EDS. Others are needed when we wish to design estimators or controllers.

3.2 Developing the discrete model

3.2.1 *Sampled-data systems*

A system with both continuous and discrete-time variables is called a *sampled-data* (or digital) system. Most of the systems that we are interested in are sampled-data systems, where a digital computer (perhaps a microprocessor) is used to control or to estimate the state of a continuous-time process. Note that in the process industry both continuous and batch processes have continuous-time variables.

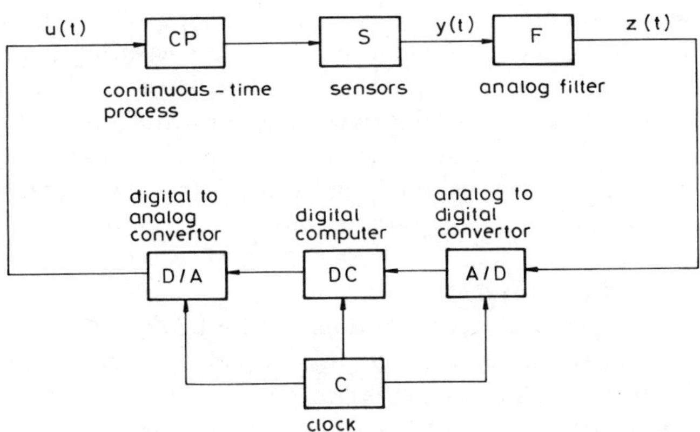

Fig. 3.1 *Sampled-data system*

Fig. 3.1 shows the most usual configuration, where the sensor produces continuous-time measurements. The presence and purpose of the analogue filter (F) needs explanation. It has already been indicated, in Section 2.2, that the process (CP + S) will contain stochastic components, some of which may include high frequencies in relation to the sampling frequency which is typically 0.03 to 1 Hz. When sampled, these high frequency signals become indistinguishable from low frequency signals – an effect known as *aliasing* – so it is important to use an analogue filter to remove the high frequency component of the sensor output. The analogue filter will be of the low-pass type: often a first-order system is used with an eigenvalue at $-2/T_s$, where T_s is the sampling interval (Smith, 1972).

The clock, which is part of the computer, sends a pulse every T_s seconds to the analogue-to-digital (A/D) convertor. The purpose of this device is to take a sample of $z(t)$ in negligible time whenever a pulse arrives. The sample taken at time KT_s is $z(KT_s)$, written z_k. Note that even though a multiplexer* is used, the time it takes to sample is negligible compared to T_s.

The same pulse which was sent by the clock to the A/D convertor also triggers the digital-to-analogue (D/A) convertor. This empties the computer output buffer of its contents, i.e. $u(KT_s)$ or u_k, and holds this value (by means of a so-called zero-order hold circuit) for the duration of the sampling interval of T_s seconds. The continuous-time process thus receives a piecewise-constant signal, $u(t)$.

One important variation on the basic scheme of Fig. 3.1 warrants discussion. It is the case when the sensor produces measurements at discrete time intervals: the measuring instrument might, for example, be an electronic weighing device or gas chromatograph. Usually there is a time delay associated with such measurements, which may range from seconds for composition analysis to hours for laboratory tests performed by human operators. In the next section we will show that, providing measurements are available at the same frequency as the A/D and D/A convertors are working, this type of sampled-data system is amenable to our analysis. Obviously, if more measuring instruments are purchased this sampling frequency can always be achieved so that pure economic factors will determine whether this or other approaches must be used. Among these other approaches are those with asynchronous or random sampling (Jury and Tyspkin, 1971; de Koning, 1980) and those using more sophisticated first and higher order hold circuits (Franklin and Powell, 1980).

3.2.2 *Equivalent discrete system*

To proceed further with the sampled-data system of Fig. 3.1 we must convert the mixture of continuous and discrete-time variables into either the continuous or the discrete time. The latter choice is convenient and moreover there are other good reasons to work with discrete-time variables. It is also necessary to account for the various modelling errors and the stochastic processes affecting the continuous-time process and sensors, as well as the imperfections of the digital computer and convertors. As we shall see, this results in a stochastic sampled-data system. The *equivalent discrete system* (*EDS*) is the discrete model having exactly the same state and output values as the stochastic sampled-data system at the sampling instants $t_0, t_1, \ldots, t_k$ etc. (Kwakernaak and Sivan, 1972). It is an extremely important concept upon which controller and estimator design procedures will be based. In this way not only the dynamics of the continuous-time process are accounted for but also the dynamics of the sampling operations. We begin by developing the EDS assuming that there are no time delays in the continuous-time system. Later this assumption is dropped.

* A multiplexer is an electronic switch enabling many signals to be sampled virtually simultaneously.

Consider first of all the analogue filter in the sampled-data system, Fig. 3.1. Recall that its purpose was to remove high-frequency stochastic components of the process: if designed as suggested in the previous section, its eigenvalue, $-2/T_s$, is considerably less than the most negative eigenvalue of the process. Therefore it is permissible to disregard the filter dynamics in our analysis of the EDS.

The A/D convertor may be represented by a switch, closing just long enough to sample the continuous system output, while the D/A convertor is modelled by such a switch followed by a hold operation. Of course, it is not essential to make a distinction between the continuous-time process and the sensors, as in Fig. 3.1; they may be lumped together in what we then choose to call the continuous(-time) system. If the sensor dynamics are significant – this will be revealed when eigenvalues are compared with those of the continuous-time process – then the dynamics can be incorporated in the dynamical equation describing the process. Neglecting non-ideal behaviour, modelling approximations, noise etc., the essentials of the sampled-data system of Fig. 3.1 can be captured in the scheme of Fig. 3.2.

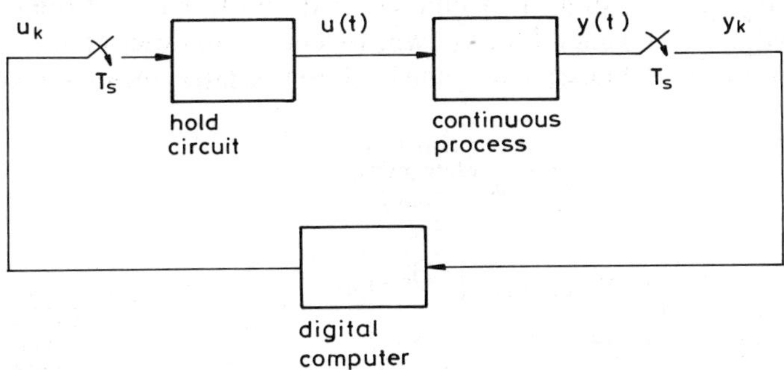

Fig. 3.2 *Schematic deterministic sampled-data system*

It is assumed that sampling occurs at the times $t_0 < t_1 < t_2$ etc. with a constant sampling interval $T_s = t_{k+1} - t_k$, $k \pm 0, 1, 2$ etc. The output of the hold element is then $u(t) = u_k$ for $t_k \leqslant t < t_{k+1}$.

The discussion so far has centred on a highly idealised situation and we must now consider which steps have to be taken to make our representation of the sampled-data system as realistic as possible. First we should account for those stochastic processes which affect the system:

disturbances (see Section 2.2.2);
sampling noise;
actuator noise.

The exact nature or cause of each random process is unlikely to be known, nor does it particularly interest us. Moreover, we have seen in Chapter 2 that continuous additive white noise can be used to model disturbances. The use of white noise to model sampling and actuator noise represents a worst-case situation implying that not only past measurements, but also mean values, are useless for characterising it. However, simplified control and estimation design procedures result from its adoption and the results of more complicated designs based on more realistic modelling by coloured noise are only marginally beneficial (Athans, 1971; Kwakernaak and Sivan, 1972).

Then there are the shortcomings associated with any attempt at mathematical modelling such as:

oversimplification of the process for modelling purposes;
inexact parameter values;
neglect of second and higher-order terms during linearisation;
neglect of the quantisation errors in convertors;
neglect of the finite word length of the computer.

It can be shown (Athans, 1971; Franklin and Powell, 1980) that these errors can be accounted for effectively by including additive white noise. Finally there is the uncertainty surrounding the value of the initial state of the dynamical equation describing the process. Since it is unlikely that all the states of the process can be measured, x_0 is, in general, unknown; we choose to describe it in terms of a random variable with known (we come back to this later) mean and covariance.

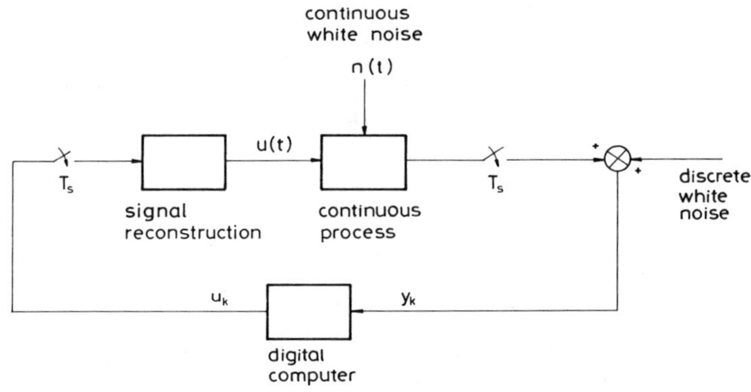

Fig. 3.3 *Schematic stochastic sampled-data system*

As far as the EDS goes as much lumping together of white noise sources as possible is desirable. This leads to the basic scheme for evaluating the EDS, shown in Fig. 3.3. Continuous additive white noise corrupts the continuous system as process (or input) noise while discrete additive white noise acts on the sampled measurement, y_k. The following result (Halyo and Caglayan, 1976) gives the EDS when the continuous-time system is the linear continuous state-space model without time delays.

Consider the sampled-data system in Fig. 3.3. The continuous stochastic dynamical system is represented by

$$\mathring{x}(t) \;=\; A(t)x(t) + B(t)u(t) + n(t) \tag{3.1a}$$

$$x(t_0) \;=\; x_0 \tag{3.1b}$$

where $n(t)$ is continuous white noise and x_0 is a random variable, and the measurements at the sampling instants are represented by

$$y_k \;=\; C_k x_k + v_k \tag{3.1c}$$

where $\{v_k\}$ is discrete white noise. Means and covariances are given by*

$$E\{x_0\} \;=\; \bar{x}_0 \tag{3.2a}$$

$$E\{(x_0 - \bar{x}_0)(x_0 - \bar{x}_0)^{\mathrm{T}}\} \;=\; P, \quad P \geqslant 0 \tag{3.2b}$$

$$E\{n(t)\} \;=\; 0 \tag{3.2c}$$

$$E\{n(t)n^{\mathrm{T}}(s)\} \;=\; N(t)\delta(t - s), \quad N(t) = 0 \tag{3.2d}$$

$$E\{v_k\} \;=\; 0 \tag{3.2e}$$

$$E\{v_k v_l^{\mathrm{T}}\} \;=\; V_k \delta_{kl}, \quad V_k \geqslant 0 \tag{3.2f}$$

and x_0, $n(t)$ and $\{v_k\}$ are mutually independent, i.e.

$$E\{(x_0 - \bar{x}_0)n^{\mathrm{T}}(t)\} \;=\; 0 \text{ for all } t \geqslant t_0 \tag{3.2g}$$

$$E\{(x_0 - \bar{x}_0)v_k^{\mathrm{T}}\} \;=\; 0 \text{ for all } k \geqslant 0 \tag{3.2h}$$

$$E\{n(t)v_k^{\mathrm{T}}\} \;=\; 0 \text{ for all } t \geqslant t_0 \text{ and } k \geqslant 0 \tag{3.2i}$$

The sample-and-hold operation is modelled by

$$u(t) \;=\; u_k, \quad t_k \leqslant t < t_{k+1} \text{ for all } k \geqslant 0 \tag{3.3}$$

where $t_k = kT_s$.

Then the EDS is the n-dimensional system

$$x_{k+1} \;=\; \Phi_k x_k + \Gamma_k u_k + w_k \tag{3.4a}$$

$$y_k \;=\; C_k x_k + v_k \tag{3.4b}$$

where x_0, $\{w_k\}$ and $\{v_k\}$ are mutually independent, $\{w_k\}$ and $\{v_k\}$ being discrete white noise with $\{v_k\}$ as above and

$$E\{w_k\} \;=\; 0 \tag{3.4c}$$

$$E\{w_k w_l^{\mathrm{T}}\} \;=\; W_k \delta_{kl} \tag{3.4d}$$

The matrices Φ_k and Γ_k are given by

$$\Phi_k \;=\; \Phi(t_{k+1}, t_k) \tag{3.5a}$$

* See Chapter 1 for definitions of the Dirac delta function $\delta(t - s)$ and the Kronecker delta δ_{kl}.

$$\Gamma_k = \int_{t_k}^{t_{k+1}} \Phi(t_{k+1}, s)B(s)\,\mathrm{d}s \tag{3.5b}$$

where $\Phi(t, \tau)$ is the transition matrix corresponding to $A(t)$, and finally

$$W_k = \int_{t_k}^{t_{k+1}} \Phi(t_{k+1}, s)N(s)\Phi^{\mathrm{T}}(t_{k+1}, s)\,\mathrm{d}s \tag{3.5c}$$

Since the output equation of the continuous-time process model is algebraic and not dynamic, it is perfectly legitimate to use eqn. 3.1*c*. Moreover, if the measurement sensor produced discrete values of the measured variables every T_s seconds (but without any appreciable time-delay), eqn. 3.1*c* would also be applicable. We see that the EDS of the linear continuous-time state-space model is the linear discrete-time state-space model (see Chapter 1) with discrete white noise processes that are additive and uncorrelated with each other. The derivation is quite easy. At the instants of time kT_s, where $k = 0, 1, 2$ etc., the continuous transition matrix $\Phi(t_{k+1}, t_k)$ corresponding to $A(t)$ has all the properties of the discrete state-space model transition matrix (Chapter 1), i.e.:

$$\Phi(t_{k+1}, t_k) = \Phi_k\Phi(t_k, t_k) = \Phi_k$$

where Φ_k is a matrix defined only at the sampling instants kT_s. Note that, Φ_k must be nonsingular for all k.

Certain simplifications are now considered. For a continuous-time process modelled by the time-invariant linear state-space representation,

$$\Phi_k = \Phi = \mathrm{e}^{AT_s} \tag{3.6a}$$

$$\Gamma_k = \Gamma = \int_0^{T_s} \mathrm{e}^{A\lambda}Bd\lambda \tag{3.6b}$$

$$W_k = W = \int_0^{T_s} \mathrm{e}^{A\lambda}N\,\mathrm{e}^{A^{\mathrm{T}}\lambda}\,\mathrm{d}\lambda \tag{3.6c}$$

These expressions underline the dependence of Φ_k, Γ_k and W_k (or Φ, Γ and W) on T_s, the sampling interval. In fact, to a first approximation, for small T_s

$$\Phi \simeq I + AT_s \tag{3.7a}$$

$$\Gamma \simeq BT_s \tag{3.7b}$$

$$W \simeq NT_s \tag{3.7c}$$

This result would also have been obtained if we had used a backwards difference approximation to $\mathring{x}(t)$ in eqn. 3.1*a*. Note that for rough calculations eqn. 3.7 is most useful. Finally we remark that as T_s becomes smaller, the EDS does not, in general, revert to the continuous-time process model since the latter does not incorporate the dynamics of sampling or signal reconstruction.

Let us now progress to examine the EDS of processes having time-delays: as in Chapter 2 a distinction is made between the various types of time delay and we consider first of all the linear continuous state-space model with input time delay.

The continuous stochastic dynamical system is represented by

$$\mathring{x}(t) = A(t)x(t) + B_0(t)u(t) + B_1(t)u(t - T) + n(t) \quad (3.8a)$$

$$x(t_0) = x_0 \quad (3.8b)$$

Measurement and sample-and-hold models are as before, eqns. 3.1c to 3.3. Then the EDS is the $(n + rT/T_s)$-dimensional system

$$z_{k+1} = \tilde{\Phi}_k z_k + \tilde{\Gamma}_k u_k + w_k \quad (3.9a)$$

$$y_k = \tilde{C}_k z_k + v_k \quad (3.9b)$$

with z_0, $\{w_k\}$, $\{v_k\}$ mutually independent and where again $\{w_k\}$ and $\{v_k\}$ are discrete white noise processes. Here,

$$\tilde{\Phi}_k = \begin{bmatrix} \Phi_k & \Gamma_k^1 & 0 \\ 0 & 0 & I_R \\ 0 & 0 & 0 \end{bmatrix}, \quad R = r(T/T_s - 1) \quad (3.10a)$$

$$\tilde{\Gamma}_k = \begin{bmatrix} \Gamma_k^0 \\ 0 \\ I_r \end{bmatrix} \quad (3.10b)$$

$$\tilde{C}_k = [C_k \quad 0 \quad 0] \quad (3.10c)$$

Proof:
The solution of eqn. 3.8a with the control, eqn. 3.3, is

$$\begin{aligned} x_{k+1} &= \Phi_k x_k + \Gamma_k^0 u_k + \int_{t_k}^{t_{k+1}} \Phi(t_{k+1}, s)B_1(s)\,\mathrm{d}s\; u_{k-T/T_s} + w_k \\ &= \Phi_k x_k + \Gamma_k^0 u_k + \Gamma_k^1 u_{k-T/T_s} + w_k \end{aligned}$$

where

$$\Gamma_k^0 = \int_{t_k}^{t_{k+1}} \Phi(t_{k+1}, s)B_0(s)\,\mathrm{d}s$$

To transform this equation into the required form we augment the state vector, x_k, with rT/T_s new states $u_{k-T/T_s}, u_{k-T/T_s+1}, \ldots, u_{k+1}$, where T/T_s is assumed to be an integer. The result follows on calling the augmented state z_k.

QED

The assumption here that T is an exact multiple of T_s is not essential and the EDS retains the same form if this is not so (see, for example, Franklin and Powell, 1980). Note that unlike Φ_k, $\tilde{\Phi}_k$ is a singular matrix.

Next we turn to the case of output time delay in our process.

The continuous stochastic dynamical system is represented by

$$\mathring{x}(t) = A(t)x(t) + B(t)u(t) + n(t) \tag{3.11a}$$

$$x(t_0) = x_0 \tag{3.11b}$$

$$y_k = C_k^0 x_k + C_k^1 x_{k-T/T_s} + v_k \tag{3.11c}$$

with means, covariances and sample-and-hold as before, eqns. 3.2 and 3.3. Then the EDS is the n-dimensional system

$$x_{k+1} = \Phi_k x_k + \Gamma_k u_k + w_k \tag{3.12a}$$

$$y_k = \tilde{C}_k x_k + v_k \tag{3.12b}$$

where x_0 and $\{w_k\}$ and $\{v_k\}$ are mutually independent, $\{w_k\}$ and $\{v_k\}$ are discrete white noise processes, Φ_k and Γ_k are given by eqn. 3.5 and

$$\tilde{C}_k = C_k^0 + C_k^1 \Phi(t_k - T, t_k) \tag{3.13}$$

This is an obvious result, following from the definition of the transition matrix (Chapter 1). Note that many sensors (e.g. gas chromatographs, titrators etc.) are realistically modelled by eqn. 3.11c.

Now the case of the continuous process model with a memoryless time delay.

The continuous stochastic dynamical system is represented by

$$\mathring{x}(t) = A(t)x(t) + B(t)u(t) + n(t) \tag{3.14a}$$

$$x(t_0) = x_0 \tag{3.14b}$$

$$y_k = C_k x_k + D_k u_{k-T/T_s} + v_k \tag{3.14c}$$

with means, covariances and sample-and-hold as before, eqns. 3.2 and 3.3. Then the EDS is the $(n + rT/T_s)$-dimensional system

$$z_{k+1} = \tilde{\Phi}_k z_k + \tilde{\Gamma}_k u_k + w_k \tag{3.15a}$$

$$y_k = \tilde{C}_k z_k + v_k \tag{3.15b}$$

where $\{w_k\}$ and $\{v_k\}$ and z_0 are mutually independent, $\{w_k\}$ and $\{v_k\}$ are discrete white noise processes and

$$\tilde{\Phi}_k = \begin{bmatrix} \Phi_k & 0 & 0 \\ 0 & 0 & I_R \\ 0 & 0 & 0 \end{bmatrix}, \quad R = r(T/T_s - 1) \tag{3.16a}$$

$$\tilde{\Gamma}_k = \begin{bmatrix} \Gamma_k \\ 0 \\ I_R \end{bmatrix} \tag{3.16b}$$

$$\tilde{C}_k = [C_k \quad D_k \quad 0]z_k \tag{3.16c}$$

The augmented state vector, z_k, is the same as that for the case where the time delay is in the input. $\tilde{\Phi}_k$ is singular.

Finally we consider the EDS of the state time-delay model. Recall from the previous chapter that the continuous-time state time-delay model is an infinite dimensional model and as such it differs from the cases considered up to now. The EDS will also be infinite dimensional. Therefore one of the approximations suggested in Section 2.2.4 must be used. If only a finite time interval, say $t_0 \leqslant t \leqslant jT$, is of interest, we can proceed as follows. Recalling eqn. 2.34:

$$\mathring{x}(t) = A(t)x(t) + \text{diag}\,[B(t - iT)u(t - iT)], \quad \text{for } i = 1, 2, 3, \ldots, j \tag{3.17}$$

it is evident that if the state vector is (further) augmented with rjT/T_s new states

$$u_{k-jT/T_s}, u_{k-jT/T_s+1}, \ldots, u_{k+1}$$

then the approach used to find the EDS for a system with input delay will work here also. The result is summarised as follows:

The continuous stochastic dynamical system is represented by

$$\mathring{x}(t) = A_0(t)x(t) + A_1(t)x(t - T) + B(t)u(t) + n(t) \tag{3.18a}$$

$$x(t_0) = x_0 \tag{3.18b}$$

Measurement and sample-and-hold models are as before, eqns. 3.1*c* to 3.3. Then, over the finite interval where $t_0 \leqslant t \leqslant jT$, the EDS is the

$$j\left(n + \frac{rT}{T_s}\right) + n$$

dimensional system

$$z_{k+1} = \tilde{\Phi}_k z_k + \tilde{\Gamma}_k u_k + w_k \tag{3.19a}$$

$$y_k = \tilde{C}_k z_k + v_k \tag{3.19b}$$

with z_0, $\{w_k\}$, $\{v_k\}$ mutually independent and where $\{w_k\}$ and $\{v_k\}$ are discrete white noise processes. Here,

$$\tilde{\Phi}_k = \begin{bmatrix} \Phi_k & \Gamma_k^j & 0 & \Gamma_k^{j-1} & \cdots & 0 \\ 0 & 0 & I_R & 0 & \cdots & 0 \\ 0 & 0 & 0 & I_R & \cdots & 0 \\ \cdot & \cdot & \cdot & \cdot & & \cdot \\ \cdot & \cdot & \cdot & \cdot & & \cdot \\ \cdot & \cdot & \cdot & \cdot & & \cdot \\ 0 & 0 & 0 & 0 & \cdots & 0 \end{bmatrix}, \quad R = r(T/T_s - 1) \tag{3.20a}$$

$$\tilde{\Gamma}_k = \begin{bmatrix} \Gamma_k^0 \\ 0 \\ \cdot \\ \cdot \\ \cdot \\ 0 \\ I_r \end{bmatrix} \quad (3.20b)$$

$$\tilde{C}_k = [C_k \quad 0 \quad . \quad . \quad . \quad 0] \quad (3.20c)$$

We conclude this section with some general remarks about the equivalent discrete system. The EDS is a model which describes not only the continuous-time process and its disturbances but also the sampling and sample-and-hold operations, which are modelled in a realistic way. At the sampling instants the EDS has exactly the same values as the stochastic sampled-data system. Between sampling instants the EDS is not defined. The EDS is so important because when we design a controller or estimator we use the EDS and not the process model. Of course, the continuous-time process is not redundant since the EDS parameters Φ_k and Γ_k are derived from it.

It has been shown that for many continuous-time processes, including those with time delays and discrete-time measuring instruments, the EDS has the same form, being the linear, discrete state-space model (see Chapter 1). The plant matrix, Φ_k, of the EDS will, in general, not be nonsingular for all k. The various means and covariances of the EDS, such as $\bar{x}_0$, W_k and V_k will not normally be known with any precision. Some of these parameters will be needed for estimator and controller design. Although it is possible to estimate these if suitable measurements are available it is extremely difficult to obtain precise values. Our philosophy is rather to use several possible values in a trial-and-error approach which culminates in a good estimator or controller response to 'worse case' conditions.

3.2.3 *Sampling period*

We must now decide on what value to choose for T_s, the interval between sampling instants. The reason is that to compute the estimate or control signal both estimator and controller design techniques will require numerical values for the EDS parameters and the EDS parameters depend on T_s (e.g. $\Phi_k = \Phi_k(T_s)$ etc.). However, the choice of T_s is not easy since the 'cost' of the control or estimate also depends on T_s. The following result (Levis, Schlueter and Athans, 1971) gives us a theoretical basis from which we can start our discussion.

Let the linear, time-invariant state-space process

$$\mathring{x}(t) = Ax(t) + Bu(t) \quad (3.21)$$

have only real eigenvalues and be controllable. Then the cost of optimally controlling the process is monotonically nondecreasing with T_s.

A similar result, one would argue heuristically, must apply to estimation so that the larger T_s is chosen to be, the poorer the estimate or control becomes (Smith, 1972). If the same conclusion can be drawn for the larger class of time-varying and time-delay processes in which we are interested, then it would always be desirable to choose T_s to be as small as possible.

Later, in Chapters 4 and 6, we will present conditions which ensure the existence and the stability of both estimators and controllers. These conditions are explicit in the EDS parameters, and are therefore dependent on T_s; the question thus arises whether an unfortunate choice of T_s could result in uncomputable or unstable estimators or controllers. It turns out that what we must ensure is basically that both controllability and reconstructibility of the process are preserved under sampling. In all but a few rather special cases we will show (see Section 3.4.3) that for any choice of T_s a controllable and reconstructible process remains so after sampling, i.e. its EDS is controllable and reconstructible. To counterbalance this reassuring result we note that the computation of the estimator or the controller will be easier (i.e. the algorithm will converge quicker) for some chosen T_s than for others – the process being, as it were, 'more controllable' or 'more reconstructible'.

At this point we must mention the implementation of the estimator or controller in a computer. With the present trend in computer hardware and software development continuing, there will be a smaller and smaller fraction of T_s used by the computer for actual computation; put another way, faster computers mean that T_s can be chosen to be smaller and smaller. A small T_s used to be associated with serious problems (such as limit cycling) owing to quantisation and roundoff – due to a finite computer wordlength – errors. However, 12 bit A/D convertors and 16 bit microprocessors have reduced these errors and should remove, or drastically reduce, these associated problems (Dorato, 1983).

Finally, the choice of T_s depends upon the process dynamics. For example, when a measuring instrument such as a gas chromatograph produces measurements only at discrete-time intervals, then it might dictate a larger T_s than would otherwise be chosen. On the other hand if a process has slow dynamics there is little point in sampling very fast. This is, in fact, roughly what Shannon's Sampling Theorem (Oliver, Pierce and Shannon, 1948) states. Based loosely on this theorem a number of rules of thumb have been suggested (Verbruggen, Peperstraete and Debruyn, 1975) to aid the process control engineer in relating the choice of T_s to the process dynamics. For example, if there is no time delay in the process (which is assumed stable with state matrix A), then

$$T_s = -\frac{1}{2\lambda_{\min}(A)} \tag{3.22}$$

while if the process contains appreciable input time delay T,

$$T_s < 1.3T \tag{3.23}$$

We illustrate the effect of the sampling interval T_s on the system dynamics by means of the following example.

Example 3.1

A first-order scalar time-invariant process

$$\dot{x}(t) = \alpha x(t) + \beta u(t) \tag{3.24}$$

is subject to proportional state feedback control by a computer:

$$u_k = -\gamma x_k \tag{3.25}$$

where $\gamma > 0$. The EDS of eqn. 3.24 is

$$x_{k+1} = e^{\alpha T_s} x_k + \frac{\beta}{\alpha}(e^{\alpha T_s} - 1)u_k \tag{3.26}$$

so that the controlled (or closed-loop) system dynamics are described by substituting eqn. 3.25 into 3.26 to give:

$$x_{k+1} = \left[\left(1 - \frac{\gamma\beta}{\alpha}\right)e^{\alpha T_s} + \frac{\gamma\beta}{\alpha}\right]x_k \tag{3.27}$$

If the controller gain, γ, has been correctly chosen the controlled system will (at least) be stable. For a choice of T_s suggested by eqn. 3.22, $0 \leqslant \gamma \leqslant -9\alpha/\beta$ to preserve stability. If we were to sample twice as fast, then we have more flexibility in choosing γ, since $0 \leqslant \gamma \leqslant -19\alpha/\beta$. If an analogue controller were used instead of the digital computer, there would be no restriction on the amount of proportional gain used. In this sense, computer control brings about a loss of stabilisability (Edgar, 1982).

3.3 Responses of simple EDS

In Section 2.2.5 the responses of a number of simple continuous models were presented. We now give the EDS of each of these models, together with its step response. Recall that the EDS has exactly the same state and output values at the sampling instants as the continuous-time process model, but that between samples it is not defined.

In Table 3.1 the EDS parameters Φ and Γ, calculated from eqns. 3.6*a* and *b* are presented together with the eigenvalues of Φ. Methods of evaluation e^{At} are given elsewhere (see, for example, Wiberg, 1971).

With Φ and Γ known it is an easy matter to simulate the response of each EDS to a unit step function (Chapter 1) applied to the input variable u_k. For clarity the effect of noise is omitted, i.e. we consider the deterministic, as opposed to the stochastic, sampled-data system of Fig. 3.2. Furthermore, the state is assumed to be initially zero; from the general EDS model, eqn. 3.4, we have therefore that

$$y_0 = 0 \tag{3.28}$$

$$y_1 = c^{\mathrm{T}}\Gamma \tag{3.29}$$

$$y_2 = c^{\mathrm{T}}\Phi\Gamma + c^{\mathrm{T}}\Gamma \tag{3.30}$$

$$y_3 = c^{\mathrm{T}}\Phi^2\Gamma + c^{\mathrm{T}}\Phi\Gamma + c^{\mathrm{T}}\Gamma \tag{3.31}$$

$$\vdots$$

$$y_k = \sum_{i=0}^{k-1} c^{\mathrm{T}}\Phi^{k-i-1}\Gamma \tag{3.32}$$

Fig. 3.4 shows the step responses.

3.4 Properties of discrete models

In this section we investigate some of the most useful properties of the EDS. Recall that the general form of the EDS is

$$x_{k+1} = \Phi_k x_k + \Gamma_k u_k + w_k \tag{3.33a}$$

$$y_k = C_k x_k + v_k \tag{3.33b}$$

where $\{w_k\}$ and $\{v_k\}$ are discrete white noise processes and x_0, $\{w_k\}$ and $\{v_k\}$ are mutually independent.

The plant matrix, Φ_k, plays a key role in determining the behaviour of eqn. 3.33. We have seen in Section 3.2.2 that Φ_k may be singular if there are time delays present in the continuous-time process. This means that eqn. 3.33*a* may not have a unique solution for $k = -1, -2, -3$ etc. However, it can be shown that, for any $k \geqslant 0$, the solution is unique whether Φ_k is singular or not, and this will be the situation of interest here. The invertibility of Φ_k has important consequences for the controllability of the EDS as will be seen subsequently.

As was the case with the linear continuous-time process models of the previous chapter, the properties such as stability of the stochastic EDS, eqn. 3.33, are the same as those of its deterministic version (i.e. with w_k and v_k identically zero in eqn. 3.33).

Note in what follows that we cannot, in general, deduce the properties of the continuous process from those of the EDS by letting T_s tend to zero. This is particularly so with controllability and reconstructibility. T_s is a (perhaps very small) positive number in our analysis; it is not zero.

3.4.1 *Mean and covariance*

In this section we derive the equations describing the dynamical behaviour of the mean and the covariance of the EDS. We also give the mean and covariance

Table 3.1

	Continuous system	$\Phi = e^{AT_s}$	$\lambda(\Phi)$	$\Gamma = \int_0^{T_s} e^{A\tau} B \, d\tau$
1	Pure gain $y(t) = \delta u(t)$	1	1	0
2	Pure integrator $\mathring{x}(t) = u(t)$	1	1	T_s
3	First order $\mathring{x}(t) = \alpha x(t) + \beta u(t)$ with $\alpha \neq 0$	$e^{\alpha T_s}$	$e^{\alpha T_s}$	$\frac{\beta}{\alpha}(e^{\alpha T_s} - 1)$
4	Second order (no zero) $\mathring{x}_1(t) = x_2(t)$ $\mathring{x}_2(t) = \alpha_1 x_1(t) + \alpha_2 x_2(t) + u(t)$			
	(i) overdamped $(-4\alpha_1 < \alpha_2^2)$	$\begin{bmatrix} \alpha_{11} & \alpha_{12} \\ \alpha_{21} & \alpha_{22} \end{bmatrix}$	$e^{\lambda_1 T_s}$ $e^{\lambda_2 T_s}$	$\begin{bmatrix} \beta_1 \\ \beta_2 \end{bmatrix}$
		$\alpha_{11} = \frac{\lambda_1}{\delta} e^{\lambda_2 T_s} - \frac{\lambda_2}{\delta} e^{\lambda_1 T_s}$	$\lambda_1 = \frac{\alpha_2 + (\alpha_2^2 + 4\alpha_1)^{\frac{1}{2}}}{2}$	$\beta_1 = \frac{1}{\delta\lambda_1} e^{\lambda_1 T_s}$
		$\alpha_{12} = \frac{1}{\delta}(e^{\lambda_1 T_s} - e^{\lambda_2 T_s})$	$\lambda_2 = \frac{\alpha_2 - (\alpha_2^2 + 4\alpha_1)^{\frac{1}{2}}}{2}$	$- \frac{1}{\delta\lambda_2} e^{\lambda_2 T_s} - \frac{1}{\alpha_1}$
		$\alpha_{21} = \frac{\alpha_1}{\delta}(e^{\lambda_1 T_s} - e^{\lambda_2 T_s})$		$\beta_2 = \frac{1}{\delta} e^{\lambda_1 T_s} - \frac{1}{\delta} e^{\lambda_2 T_s}$
		$\alpha_{22} = \frac{\lambda_1}{\delta} e^{\lambda_1 T_s} - \frac{\lambda_2}{\delta} e^{\lambda_2 T_s}$		
	where	$\delta = (\alpha_2^2 + 4\alpha_1)^{\frac{1}{2}}$		

(ii) critically damped ($4\alpha_1 = \alpha_2^2$)

$$\begin{bmatrix} \alpha_{11} & \alpha_{12} \\ \alpha_{21} & \alpha_{22} \end{bmatrix} \qquad \begin{matrix} e^{\lambda T_s} \\ e^{\lambda T_s} \end{matrix} \qquad \begin{bmatrix} \beta_1 \\ \beta_2 \end{bmatrix}$$

$$\alpha_{11} = (1 - \lambda T_s)\, e^{\lambda T_s}$$

$$\alpha_{12} = T_s\, e^{\lambda T_s}$$

$$\alpha_{21} = \alpha_1 T_s\, e^{\lambda T_s}$$

$$\alpha_{22} = \left(1 + \frac{\alpha_2}{2} T_s\right) e^{\lambda T_s}$$

where

$$\lambda = \frac{\alpha_2}{2}$$

$$\beta_1 = \left(\frac{2\alpha_2 T_s - 4}{\alpha_2^2}\right) e^{\lambda T_s} - \frac{1}{\alpha_1}$$

$$\beta_2 = T_s\, e^{\lambda T_s}$$

(iii) underdamped ($-4\alpha_1 > \alpha_2^2$)

$$\begin{bmatrix} \alpha_{11} & \alpha_{12} \\ \alpha_{21} & \alpha_{22} \end{bmatrix} \qquad \begin{matrix} e^{\left(\frac{\alpha_2}{2} + j\gamma\right)T_s} \\ e^{\left(\frac{\alpha_2}{2} - j\gamma\right)T_s} \end{matrix} \qquad \begin{bmatrix} \beta_1 \\ \beta_2 \end{bmatrix}$$

$$\alpha_{11} = e^{\frac{\alpha_2 T_s}{2}} \left[\cos \gamma T_s - \frac{\alpha_2}{2\gamma} \sin \gamma T_s\right]$$

$$\alpha_{12} = \frac{e^{\frac{\alpha_2 T_s}{2}} \sin \gamma T_s}{\gamma}$$

$$\alpha_{21} = \frac{\alpha_1\, e^{\frac{\alpha_2 T_s}{2}} \sin \gamma T_s}{\gamma}$$

$$\alpha_{22} = e^{\frac{\alpha_2 T_s}{2}} \left[\cos \gamma T_s + \frac{\alpha_2}{2\gamma} \sin \gamma T_s\right]$$

where

$$\gamma = \left(-\alpha_1 - \frac{\alpha_2^2}{4}\right)^{\frac{1}{2}}$$

$$\beta_1 = \frac{e^{\frac{\alpha_2 T_s}{2}}}{\alpha_1} \left[\cos \gamma T_s - \frac{\alpha_2}{2\gamma} \sin \gamma T_s\right]$$

$$\beta_2 = \frac{1}{\gamma}\, e^{\frac{\alpha_2 T_s}{2}} \sin \gamma T_s$$

Table 3.1 *Continued*

	Continuous system	$\Phi = e^{AT_s}$	$\lambda(\Phi)$	$\Gamma = \int_0^{T_s} e^{A\tau}B\,d\tau$
5	General second order		Same as second order (no zero)	
6	Memoryless time delay $y(t) = u(t - T)$ $T = 2T_s$	$\begin{bmatrix} 1 & 0 & 0 \\ 0 & 0 & 1 \\ 0 & 0 & 0 \end{bmatrix}$	0 0 1	$\begin{bmatrix} 0 \\ 0 \\ 1 \end{bmatrix}$
7	Input time delay $\mathring{x}(t) = \alpha x(t) + \beta u(t - T)$ $T = 2T_s$	$\begin{bmatrix} e^{\alpha T_s} & \frac{\beta}{\alpha}(e^{\alpha T_s} - 1) & 0 \\ 0 & 0 & 1 \\ 0 & 0 & 0 \end{bmatrix}$	0 0 $e^{\alpha T_s}$	$\begin{bmatrix} 0 \\ 0 \\ 1 \end{bmatrix}$
8	Output time delay $\mathring{x}(t) = \alpha x(t) + \beta u(t)$ $y(t) = x(t - T)$ $T = 2T_s$	$e^{\alpha T_s}$	$e^{\alpha T_s}$	$\frac{\beta}{\alpha}(e^{\alpha T_s} - 1)$

9 State time delay

$$\dot{x}(t) = \alpha_0 x(t) + \alpha_1 x(t - T) + \beta u(t)$$

$$t_0 \leqslant t \leqslant 2T$$

$$T = 5T_s$$

$$\begin{bmatrix} \alpha_{11} & \alpha_{12} & \alpha_{13} & 0 & 0 & 0 & 0 & 0 & 0 & 0 & 0 & 0 & 0 \\ 0 & \alpha_{22} & \alpha_{23} & 0 & 0 & 0 & 0 & 0 & \beta_1 & 0 & 0 & 0 & 0 \\ 0 & 0 & \alpha_{33} & \beta_1 & 0 & 0 & 0 & 0 & 0 & 0 & 0 & 0 & 0 \\ 0 & 0 & 0 & 0 & 1 & 0 & 0 & 0 & 0 & 0 & 0 & 0 & 0 \\ 0 & 0 & 0 & 0 & 0 & 1 & 0 & 0 & 0 & 0 & 0 & 0 & 0 \\ 0 & 0 & 0 & 0 & 0 & 0 & 1 & 0 & 0 & 0 & 0 & 0 & 0 \\ 0 & 0 & 0 & 0 & 0 & 0 & 0 & 1 & 0 & 0 & 0 & 0 & 0 \\ 0 & 0 & 0 & 0 & 0 & 0 & 0 & 0 & 1 & 0 & 0 & 0 & 0 \\ 0 & 0 & 0 & 0 & 0 & 0 & 0 & 0 & 0 & 1 & 0 & 0 & 0 \\ 0 & 0 & 0 & 0 & 0 & 0 & 0 & 0 & 0 & 0 & 1 & 0 & 0 \\ 0 & 0 & 0 & 0 & 0 & 0 & 0 & 0 & 0 & 0 & 0 & 1 & 0 \\ 0 & 0 & 0 & 0 & 0 & 0 & 0 & 0 & 0 & 0 & 0 & 0 & 1 \\ 0 & 0 & 0 & 0 & 0 & 0 & 0 & 0 & 0 & 0 & 0 & 0 & 0 \end{bmatrix} \quad \begin{bmatrix} 0 \\ 0 \\ 0 \\ 0 \\ 0 \\ 0 \\ 0 \\ 0 \\ 0 \\ 0 \\ \alpha_{11} \\ \alpha_{11} \\ \alpha_{11} \end{bmatrix} \quad \begin{bmatrix} \beta_1 \\ 0 \\ 0 \\ 0 \\ 0 \\ 0 \\ 0 \\ 0 \\ 0 \\ 0 \\ 0 \\ 0 \\ 1 \end{bmatrix}$$

$$\alpha_{11} = \alpha_{22} = \alpha_{23} = e^{\alpha_0 T_s}$$

$$\alpha_{12} = \alpha_{23} = \alpha_1 T_s e^{\alpha_0 T_s}$$

$$\alpha_{13} = \frac{\alpha_1^2 T_s^2}{2} e^{\alpha_0 T_s}$$

$$\beta_1 = \frac{\beta}{\alpha_0}(e^{\alpha_0 T_s} - 1)$$

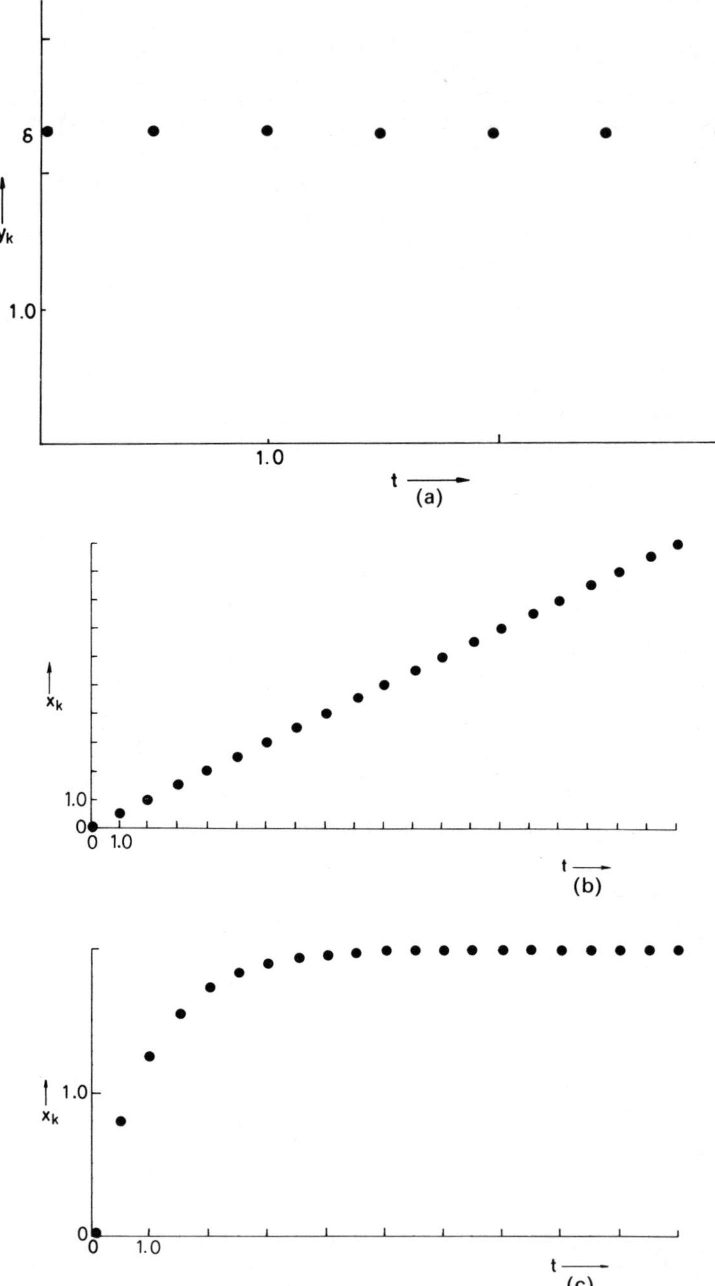

Fig. 3.4 *a* Pure gain
b Pure integrator
c First order, $\alpha < 0$

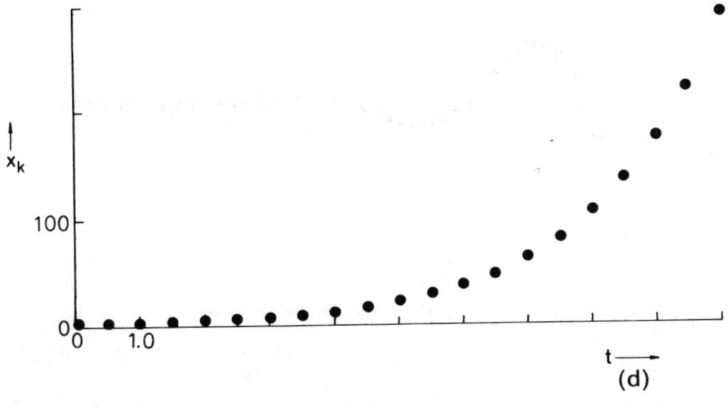

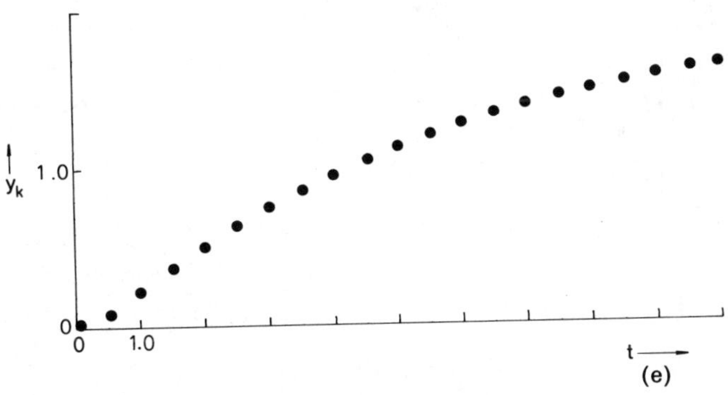

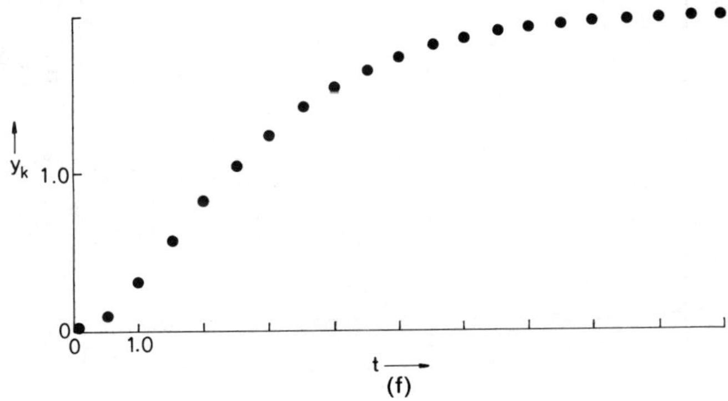

Fig. 3.4 *Continued*

d First order, $\alpha > 0$

e Second order, no zero, overdamped, $-4\alpha_1 < \alpha_2^2$, $\alpha_1 < 0$ and $\alpha_2 < 0$

f Second order, no zero, critically damped, $-4\alpha_1 = \alpha_2^2$, $\alpha_2 < 0$

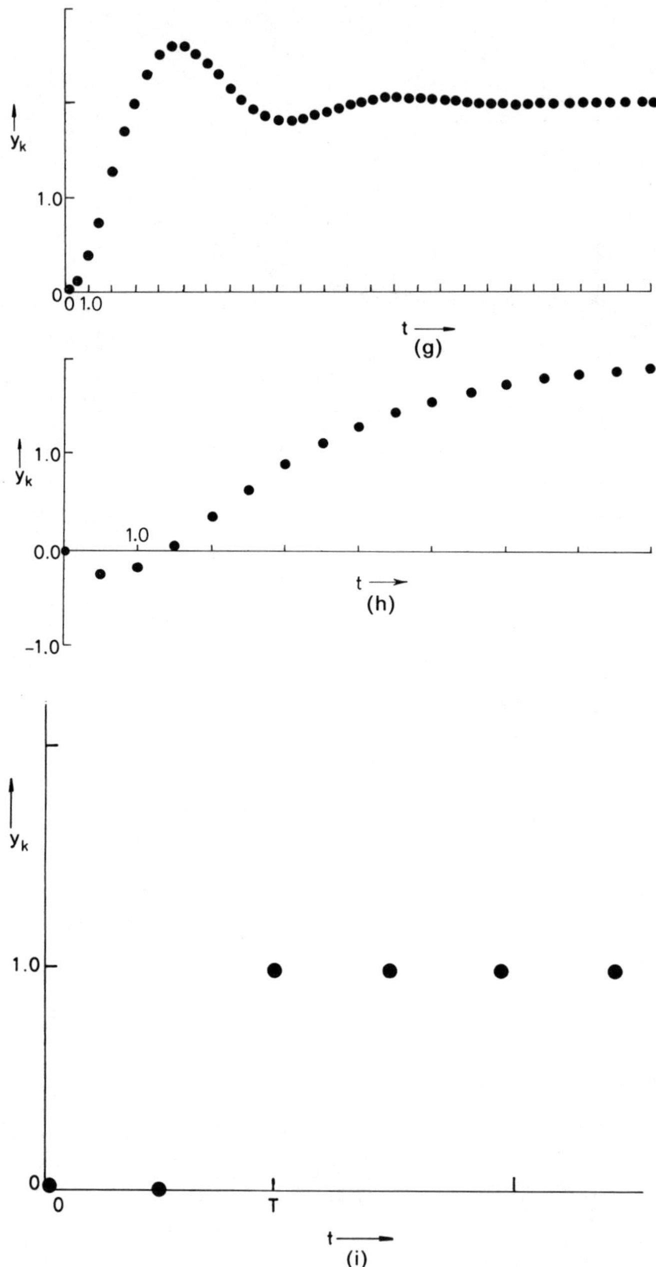

Fig. 3.4 *Continued*

g Second order, no zero, underdamped, $-4\alpha_1 > 4\alpha_2^2$, $\alpha_1 < 0$, $\alpha_2 < 0$

h General second order, inverse response, $\alpha_1 < 0$, $\alpha_2^2 = -4\alpha_1$, $\gamma_1 = 1$, $\gamma_2 = -1$

i Memoryless time delay

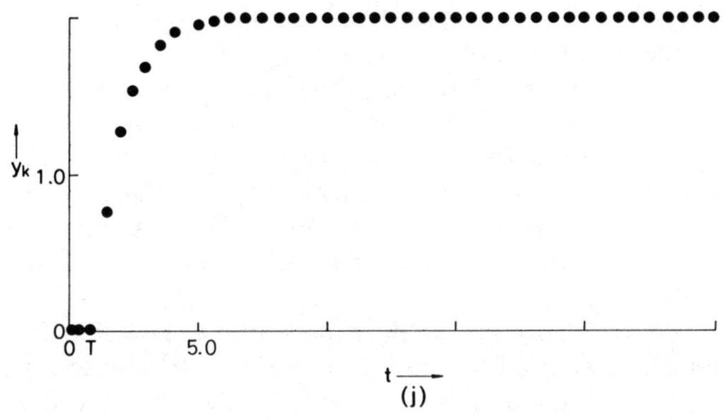

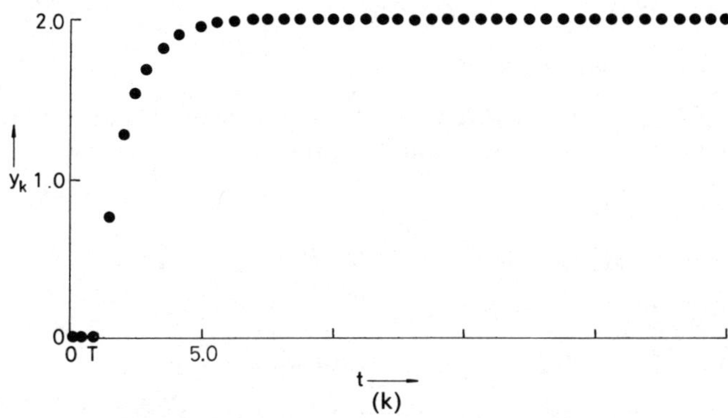

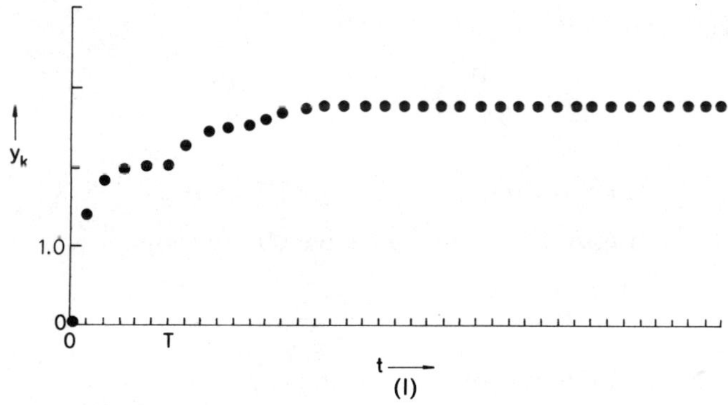

Fig. 3.4 *Continued*

j Input time delay, $\alpha < 0$
k Output time delay, $\alpha < 0$
l State time delay, $-\alpha_0 > \alpha_1 \geqslant 0$

of the continuous-time process model between the sampling instants as a function of the values at the sampling instants.

We will usually only be interested in the first two moments, the mean and covariance (see Chapter 1), of the system state and output. If the additive noise terms and x_0 are all normally distributed (see Chapter 1) then the state, x_k, of the EDS is also normal, so that the mean and covariance of x_k completely describe the statistical behaviour of the EDS. The Gaussian assumption is justified in a number of situations encountered in process control (Aström, 1970).

For the process control engineer the expressions we give shortly for the mean and covariance can be used to check the results of (closed-loop) simulations. This explains why no control term, u_k, is to be found in the EDS, since it is assumed that state feedback is used.

Consider the EDS

$$x_{k+1} = \Phi_k x_k + w_k \tag{3.34}$$

where $\Phi_k \triangleq \Phi(t_{k+1}, t_k)$ and $\Phi(t, \tau)$ is the state transition matrix corresponding to $A(t)$. Furthermore, the means and covariances are given by

$$E\{x_0\} = \bar{x}_0 \tag{3.35a}$$

$$E\{(x_0 - \bar{x}_0)(x_0 - \bar{x}_0)^T\} = P, \quad P \geqslant 0 \tag{3.35b}$$

$$E\{w_k\} = 0 \tag{3.35c}$$

$$E\{w_k w_l^T\} = W_k \delta_{kl}, \quad W_k \geqslant 0 \text{ for all } k \tag{3.35d}$$

Then at the sampling instants the mean of x_k is given by

$$E\{x_k\} \triangleq \bar{x}_k = \Phi(t_k, t_0)\bar{x}_0 \tag{3.36}$$

and the covariance

$$E\{(x_k - \bar{x}_k)(x_l - \bar{x}_l)^T\} \triangleq \Sigma_{k,l}$$

is given by

$$\Sigma_{k,l} = \Sigma_{k,k}\Phi^T(t_l, t_k) \tag{3.37a}$$

where $\Sigma_{k,k} \triangleq \Sigma_k$ is found from the recursive relationship

$$\Sigma_{k+1} = \Phi_k \Sigma_k \Phi_k^T + W_k \tag{3.37b}$$

with $\Sigma_0 = P$.

Between the sampling instants the mean is

$$E\{x(t)\} = \Phi(t, t_k)\bar{x}_k \tag{3.38}$$

and covariance

$$E\{(x(t) - E\{x(t)\})(x(t) - E\{x(t)\})^T\}$$

$$= \Phi(t, t_k)\Sigma_k\Phi^{\mathrm{T}}(t, t_k) + \int_{t_k}^{t} \Phi(t, \tau)N(\tau)\Phi^{\mathrm{T}}(t, \tau)\,\mathrm{d}\tau \qquad (3.39)$$

where

$$E\{n(t)n^{\mathrm{T}}(s)\} = N(t)\delta(t - s)$$

Proof:
At the sampling instants, from eqns. 3.34 and 3.35*c*

$$E\{x_{k+1}\} = \Phi_k E\{x_k\}$$

or

$$\begin{aligned} E\{x_k\} &= \Phi\{t_k, t_{k-1}\}E\{x_{k-1}\} \\ &= \Phi(t_k, t_{k-1})\Phi(t_{k-1}, t_{k-2}) \ldots \Phi(t_1, t_0)E\{x_0\} \\ &= \Phi(t_k, t_0)\bar{x}_0 \end{aligned}$$

from the transition property of $\Phi(t, \tau)$, see Chapter 1. Using eqns. 3.34, 3.36 and the same transition matrix property we have that

$$\Sigma_{k+1} = E\{(\Phi_k x_k + w_k - \Phi_k\bar{x}_k)(\Phi_k x_k + w_k - \Phi_k\bar{x}_k)^{\mathrm{T}}\}$$

Since eqn. 3.34 is an EDS, x_0 and $\{w_k\}$ are mutually independent (Section 3.2.2) so that after some manipulation

$$\Sigma_{k+1} = E\{\Phi_k(x_k - \bar{x}_k)(x_k - \bar{x}_k)^{\mathrm{T}}\Phi_k^{\mathrm{T}}\} + W_k$$

and eqn. 3.37 follows. Since $x_l = \Phi(t_l, t_k)x_k$ it is easy to show that

$$\Sigma_{k,l} = \Sigma_{k,k}\Phi^{\mathrm{T}}(t_l, t_k) = \Sigma_k\Phi^{\mathrm{T}}(t_l, t_k)$$

Now consider the evolution of the mean between the sampling instants. The state is determined by

$$\mathring{x}(t) = A(t)x(t) + n(t)$$

or

$$x(t) = \Phi(t, t_k)\bar{x}_k + \int_{t_k}^{t} \Phi(t, \tau)n(\tau)\,\mathrm{d}\tau$$

where $n(\ )$ is continuous white noise, with zero mean. Taking expectations of this last equation yields eqn. 3.38. Finally, the covariance is given by

$$\begin{aligned} &E\{(x(t) - \Phi(t, t_k)\bar{x}_k)(x(t) - \Phi(t, t_k)\bar{x}_k)^{\mathrm{T}}\} \\ &\quad = \Phi(t, t_k)E\{x_k x_k^{\mathrm{T}} - x_k\bar{x}_k^{\mathrm{T}} - \bar{x}_k x_k^{\mathrm{T}} + \bar{x}_k\bar{x}_k^{\mathrm{T}}\}\Phi^{\mathrm{T}}(t, t_k) \\ &\qquad + E\left\{\int_{t_k}^{t} \Phi(t, \tau)n(\tau)\,\mathrm{d}\tau\left[\int_{t_k}^{t} \Phi(t, \tau)n(\tau)\,\mathrm{d}\tau\right]^{\mathrm{T}}\right\} \end{aligned}$$

Eqn. 3.39 follows on noting that the last term may be considered to be an integral multiplied by a constant factor, which may therefore be considered to be a part of the integrand.

The foregoing could be generalised to include an EDS with control term, u_k, without difficulty. An important special case of the result is when $\Phi_k = \Phi = e^{AT_s}$ (eqn. 3.6*a*); at the sampling instants the mean and covariance are given, respectively, by

$$\bar{x}_k = e^{AkT_s}\bar{x}_0 \tag{3.40}$$

$$\Sigma_{k+1} = e^{AT_s}\Sigma_k \, e^{A^{\mathrm{T}}T_s} + W_k \tag{3.41}$$

where again $\Sigma_0 = P$. If the initial condition of the state is completely known (so $P = 0$) and moreover if the noise is stationary ($W_k = W$) then

$$\bar{x}_k = e^{AkT_s}x_0 \tag{3.42}$$

$$\Sigma_{k+1} = e^{AkT_s}W \, e^{A^{\mathrm{T}}kT_s} + \Sigma_k \tag{3.43}$$

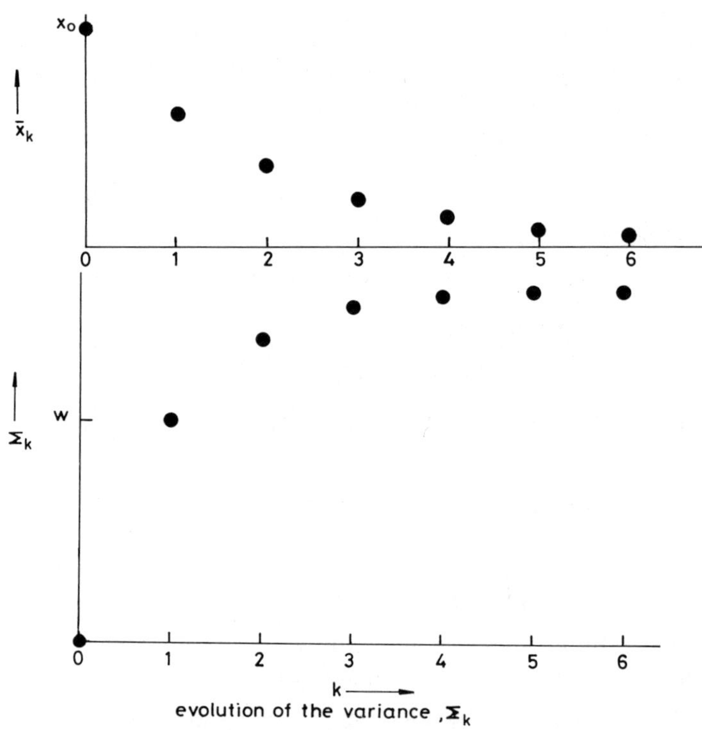

Fig. 3.5 *Statistical properties of a first-order s.i.s.o. sampled process with deterministic initial state*
a Evolution of the mean, $\bar{x}_k$
b Evolution of the variance, Σ_k

Fig. 3.5 shows how the mean and covariance evolve when a s.i.s.o. stable first-order continuous process is sampled. Another interesting case is when there is no noise ($W_k = 0$ for all k) but the initial condition x_0 is a stochastic variable.

Eqns. 3.40 and 3.41 then become

$$\bar{x}_k = e^{AkT_s}\bar{x}_0 \tag{3.44}$$

$$\Sigma_k = e^{AkT_s}P\,e^{A^TkT_s} \tag{3.45}$$

Since it has been assumed that the white noise sequence $\{w_k\}$ has zero mean, eqns. 3.40 and 3.44 are the same. Fig. 3.6 illustrates the solution of eqns. 3.44 and 3.45 for a s.i.s.o. stable continuous first-order sampled process.

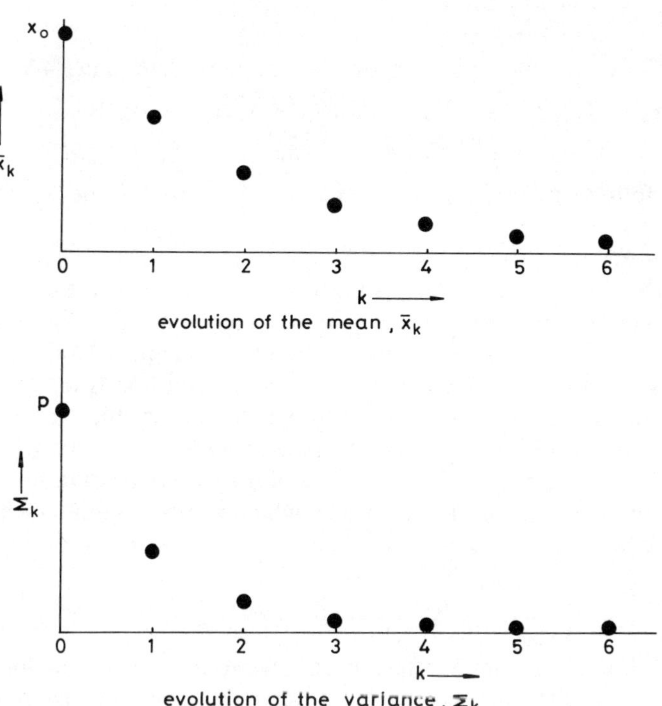

Fig. 3.6 *Statistical properties of a first-order s.i.s.o. sampled process with no additive noises*
a Evolution of the mean, $\bar{x}_k$
b Evolution of the variance, Σ_k

Finally, we must consider the mean and covariance of the output of the EDS.

Let the EDS be, as before, eqns. 3.34 and 3.35 but with the output given by

$$y_k = C_k x_k + v_k \tag{3.46}$$

where

$$E\{v_k\} = 0 \tag{3.47a}$$

and

$$E\{v_k v_k^{\mathrm{T}}\} = V_k \text{ for all } k \qquad (3.47b)$$

Then at the sampling instants the mean of the output is given by

$$E\{y_k\} \triangleq \bar{y}_k = C_k \bar{x}_k \qquad (3.48)$$

and

$$E\{(y_k - \bar{y}_k)(y_k - \bar{y}_k)^{\mathrm{T}}\} = C_k \Sigma_k C_k^{\mathrm{T}} + V_k \qquad (3.49)$$

where $\bar{x}_k$ and Σ_k are as given before.

Proof:
The proof of eqn. 3.48 follows at once from eqns. 3.46 and 3.47*a*. Now

$$(y_k - \bar{y}_k)(y_k - \bar{y}_k)^{\mathrm{T}} = C_k(x_k - \bar{x}_k)(x_k - \bar{x}_k)^{\mathrm{T}} C_k^{\mathrm{T}} - C_k(x_k - \bar{x}_k)v_k^{\mathrm{T}} - v_k(x_k - \bar{x}_k)^{\mathrm{T}} C_k^{\mathrm{T}} + v_k v_k^{\mathrm{T}}$$

Eqn. 3.49 follows from the properties of the EDS on taking expectations.

3.4.2 *Stability*

Since the EDS is a linear model it is permissible to talk about the stability of the EDS. The stability (for our purposes uniform asymptotic stability) of the EDS is defined in exactly the same way as for a continuous linear model of Chapter 2. The steady-state of the EDS and its associated continuous process model are obviously the same. If an EDS is unstable then so is the continuous model. However, if the continuous model is unstable it may not be so obvious from the behaviour of the state of the EDS. Fig. 3.7 illustrates this point, not only for the case of a time-varying continuous model which exhibits a finite escape time but also for the well-known harmonic oscillator.

$$\mathring{x}(t) = \begin{bmatrix} 0 & 1 \\ -1 & 0 \end{bmatrix} x(t) \qquad (3.50)$$

The instability does not manifest itself owing to a phenomenon known as *aliasing*. The oscillations which occur between sampling instants are called hidden oscillations or ripple.

There are occasions when it is either necessary or desirable to check the stability of an EDS without reference to the associated continuous model. For a time-varying EDS there is at present no result based upon the matrix measure which corresponds to that of the continuous, time-varying situation. The following result (Kalman and Bertram, 1960; Wiberg, 1971) is generally too conservative to be of much practical use.

Given the unforced EDS

$$x_{k+1} = \Phi_k x_k$$

Then if $\|\Phi_k\| < 1$ for all k the null solution $x = 0$ is uniformly asymptotically stable.

The hyperstability theorem of Popov (Landau, 1979) is a much stronger result. We assume that the unforced EDS can be formulated as a feedback system, see Fig. 3.8, such that

$$z_{k+1} = Fz_k + Gu_k \tag{3.51a}$$

$$x_k = Cz_k + Du_k \tag{3.51b}$$

$$u_k = -s_k \tag{3.51c}$$

$$s_k = T_k x_k \tag{3.51d}$$

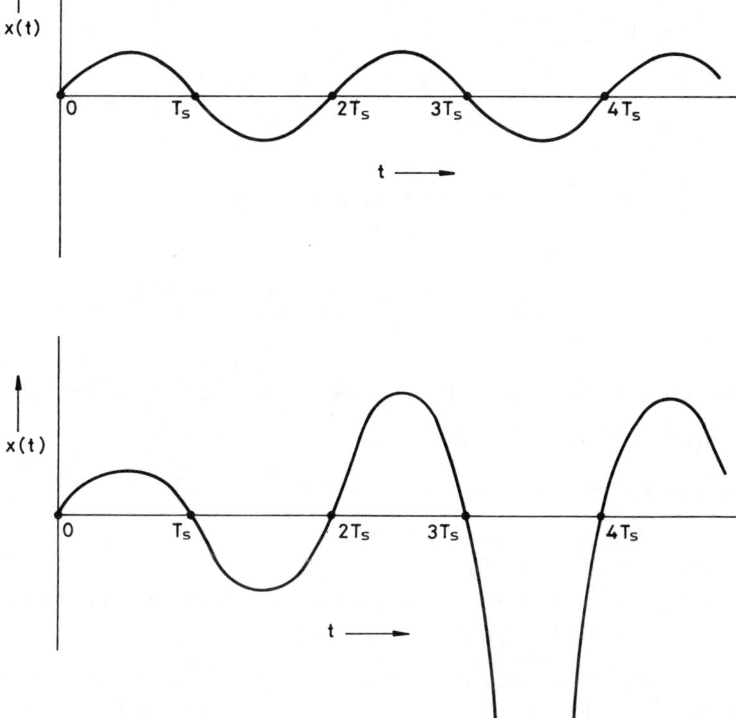

Fig. 3.7 *Intersampling phenomena*

Note that all time-varying parameters are collected in T_k in the feedback block. Eqn. 3.51 must fulfil the Popov inequality

$$\sum_{k=0}^{k_1} x_k^T s_k \geqslant \alpha \tag{3.52}$$

for all k_1, where α is a negative constant independent of k_1. Now if the time-invariant subsystem of Fig. 3.8, i.e. eqns. 3.51*a* to 3.51*c* is such that the EDS $x_{k+1} = \Phi_k x_k$ is uniformly asymptotically stable for all T_k satisfying eqn. 3.52 then the EDS is hyperstable, and the time-invariant subsystem is a hyperstable subsystem. The following result gives a condition to ensure hyperstability (Landau, 1979).

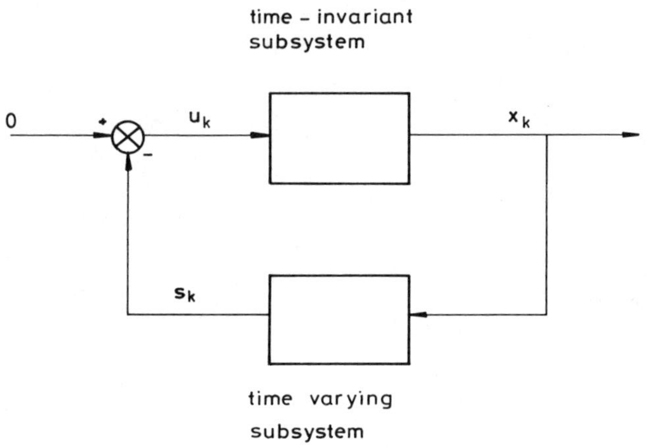

Fig. 3.8 *Unforced feedback system equivalent to the EDS* $x_{k+1} = \Phi_k x_k$

The EDS described by eqn. 3.51 is hyperstable if and only if the matrix

$$H(\sigma) = C(\sigma I - F)^{-1}G$$

is strictly positive real, i.e.

(i) $H(\sigma)$ is real for all real σ.
(ii) $|\{\lambda_i(H)\}| < 1$ for all i.
(iii) $H(e^{j\omega}) + H^T(e^{-j\omega})$ is a positive definite Hermitian† matrix for all real values of ω.

As with continuous systems, the stability of a time-invariant EDS is much better understood. Recall that a continuous unforced model was uniformly asymptotically stable if the eigenvalues, $\lambda_i(A)$, all had negative real parts. Now if the continuous model is stable, so the corresponding EDS will be stable. What about its eigenvalues?

† The matrix H is Hermitian if

$$H(\sigma) = H^T(\sigma^*)$$

where σ^* is the complex conjugate of σ.

$$\begin{aligned}
\lambda_i(\Phi) &= \lambda_i(\mathrm{e}^{AT_s}) \\
&= \mathrm{e}^{\lambda_i(A)T_s} \\
&= \mathrm{e}^{(\alpha+j\beta)T_s}, \text{ say} \\
&= \mathrm{e}^{\alpha T_s}\mathrm{e}^{j\beta T_s} \\
&= \mathrm{e}^{\alpha T_s}(\cos \beta T_s + j \sin \beta T_s)
\end{aligned}$$

If Re $\lambda_i(A) = \alpha$ is negative then we see that all the eigenvalue of Φ are constrained to lie within a circle of radius one, at the centre of the complex plane. Actually this turns out to be not only sufficient but necessary for asymptotic stability.

Given the unforced, time-invariant EDS

$$x_{k+1} = \Phi x_k$$

Then if and only if, for $i = 1, 2, \ldots, n$

$$|\lambda_i(\Phi)| < 1$$

the EDS is uniformly asymptotically stable.

Fig. 3.9 contrasts the stability domains in the continuous and discrete cases.

Finally we recall that the EDS of the state time-delay model, Section 3.2.2, was valid only over a finite time interval and is therefore not amenable to investigations concerning stability. For this situation the following result (Mori *et al.*, 1982), is useful.

Given the state time-delay model

$$x_{k+1} = \Phi x_k + \psi x_{k-T}$$

where Φ and ψ are $(n \times n)$ time-invariant matrices then if

$$1 - \|\Phi\| > \|\psi\|$$

the model is asymptotically stable.

3.4.3 *Controllability, reachability, reconstructibility and observability*

We recall once more for ease of reference the general form of the EDS:

$$x_{k+1} = \Phi_k x_k + \Gamma_k u_k + w_k \tag{3.33a}$$

$$y_k = C_k x_k + v_k \tag{3.33b}$$

Now in Section 2.3.4 the uniform complete controllability of the continuous-time model has been shown to be equivalent to the existence of a uniformly bounded control which transfers some initial state to zero in a finite time. The symmetric controllability Gramian was defined to be

$$W(t_0, t_1) \triangleq \int_{t_0}^{t_1} \Phi(t_0, s)B(s)B^{\mathrm{T}}(s)\Phi^{\mathrm{T}}(t_0, s)\,\mathrm{d}s \tag{2.56}$$

Recall from Section 3.2.2 that $\Phi_k = \Phi(k+1, k)$ is the discrete state transition matrix and that (with $w_k \equiv 0$) the solution of eqn. 3.33*a* for $l > k$ is

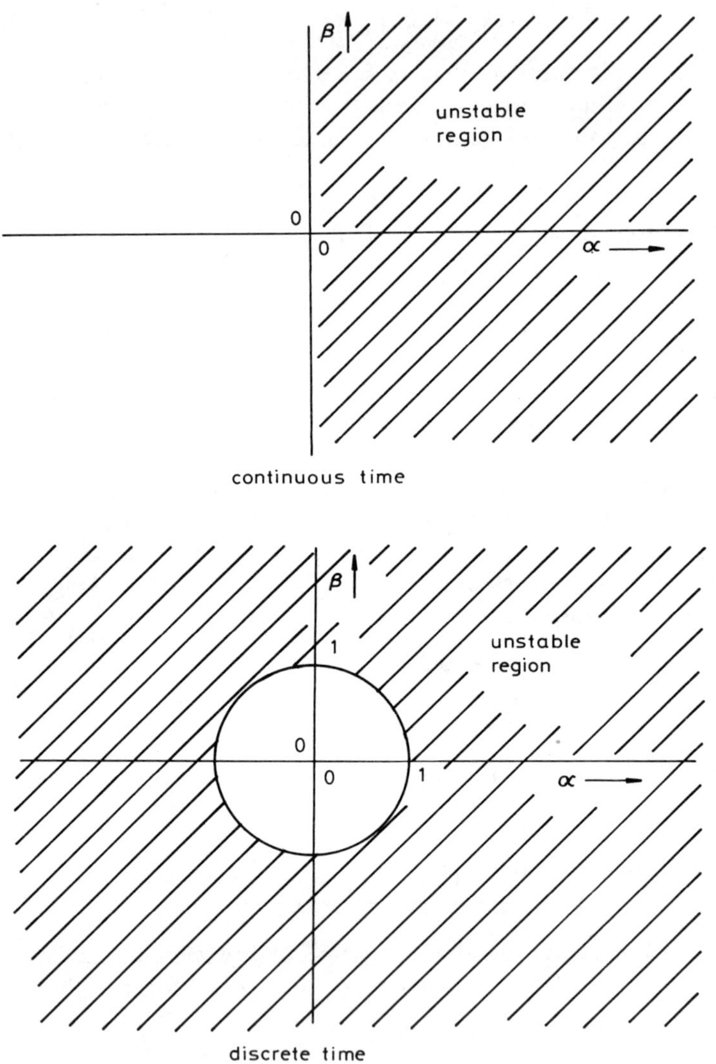

Fig. 3.9 *Argand diagrams of the complex plane showing stable and unstable regions for the eigenvalue* $\lambda = \alpha + j\beta$

a Continuous time

b Discrete time

$$x_l = \Phi(l, k)x_k + \sum_{i=k}^{l-1} \Phi(l, i+1)\Gamma_i u_i \tag{3.53}$$

Now with $x_k = x_0$ the initial state, and $x_l = 0$ the terminal state, it is obvious that unless $\Phi(l, k)$ is invertible we can never find a suitable control (let alone a uniformly bounded one). We have already seen in Section 3.2.2 that the EDS of a continuous process containing a time delay has a singular matrix Φ_k and so we can expect the concept of (uniform complete) controllability for sampled-data systems to be of limited practical use.

The EDS, eqn. 3.33, is not uniformly completely controllable if Φ_k is singular for any k.

Note that the problem did not arise in the case of continuous-time systems since $\Phi^{-1}(t, t_0)$ always exists.

The above dilemma is resolved by introducing the concept of *uniform complete reachability*. This is the existence of a uniformly bounded control which transfers the system from zero to any terminal state in a finite time. With $x_k = 0$ and letting $x_l = x$, any chosen final state, eqn. 3.53 may be written

$$x = \sum_{i=k}^{l-1} \Phi(l, i+1)\Gamma_i u_i$$

$$= [\Gamma_{l-1}\Phi(l, l-1)\Gamma_{l-2} \ldots \Phi(l, k+1)\Gamma_k] \begin{bmatrix} u_{l-1} \\ u_{l-2} \\ \cdot \\ \cdot \\ \cdot \\ u_k \end{bmatrix} \tag{3.54a}$$

$$\triangleq C_{k,l} \begin{bmatrix} u_{l-1} \\ \cdot \\ \cdot \\ \cdot \\ u_k \end{bmatrix} \tag{3.54b}$$

Clearly, the system is reachable if and only if the rank of $C_{k,l}$ is n. Defining now the *discrete reachability Gramian* as

$$\tilde{W}(k, l) \triangleq C_{k,l}C_{k,l}^{\mathrm{T}}$$

$$= \sum_{i=k}^{l-1} \Phi(l, i+1)\Gamma_i\Gamma_i^{\mathrm{T}}\Phi^{\mathrm{T}}(l, i+1) \tag{3.55}$$

we see that the system is reachable if and only if $\tilde{W}(k, l) > 0$. The crucial difference between the controllability and reachability Gramians is that the latter uses only the matrices $\Phi(k + 1, k) = \Phi_k$ and not their inverses (which need not, therefore, necessarily exist). Reachability is clearly a more practical concept than controllability.

The test for *discrete-time uniform complete reachability* is similar to that given in Section 2.3.4 for the continuous-time case. We quote the following result (Kwakernaak and Sivan, 1972).

The EDS, eqn. 3.33, is uniformly completely reachable if there exists an integer $j > 0$ and positive scalars $\alpha_0(j)$, $\alpha_1(j)$, $\beta_0(j)$ and $\beta_1(j)$ such that:

1. $\tilde{W}(k, k + j) > 0$ for all k
2. $\alpha_0(j)I \leqslant \tilde{W}^{-1}(k, k + j) \leqslant \alpha_1(j)I$ for all k
3. $\beta_0(j)I \leqslant \Phi^T(k + j, k)\tilde{W}^{-1}(k, k + j)\Phi(k + j, k) \leqslant \beta_1(j)I$ for all k

---

Note that for continuous-time systems reachability implies and is implied by controllability. For discrete-time systems reachability always implies controllability but only if the inverse of the plant matrix, Φ_k, of the EDS exists does controllability imply reachability.

Just as with continuous-time systems there is a *duality* between the properties of discrete-time systems. The duality is between:

reachability and observability;

controllability and reconstructibility,

the same as discussed in Chapter 2. To avoid the existence problems occurring with the reconstructibility Gramian, we proceed straightaway to the discrete-time observability Gramian, defined as the dual of the reachability Gramian and thus given by

$$\tilde{M}(k, l) = \sum_{i=k}^{l-1} \Phi^T(i, k)C_i^T C_i \Phi(i, k) \tag{3.56}$$

Because of its importance we state the test for *discrete-time uniform complete observability*.

The EDS, eqn. 3.33, is uniformly completely observable if there exists an integer $j > 0$ and positive scalars $\alpha_0(j)$, $\alpha_1(j)$, $\beta_0(j)$ and $\beta_1(j)$ such that:

1. $\tilde{M}(k, k + j) > 0$ for all k
2. $\alpha_0(j)I \leqslant \tilde{M}^{-1}(k, k + j) \leqslant \alpha_1(j)I$ for all k
3. $\beta_0(j)I \leqslant \Phi(k + j, k)\tilde{M}^{-1}(k, k + j)\Phi^T(k + j, k) \leqslant \beta_1(j)I$ for all k

---

For continuous systems observability implies and is implied by reconstructibility. For discrete systems, observability always implies reconstructibility, but only if the inverse of the plant matrix, Φ_k, of the EDS exists does reconstructibility imply observability.

Weaker, or less stringent, conditions than reachability (or observability) are those of stabilisability (detectability) discussed in the next section.

We now turn to the effect that sampling has upon reachability and observability. It has already been pointed out that letting T_s tend to zero can lead to erroneous conclusions regarding the reachability or observability of the continuous-time process, and for obvious reasons. See Example 3.2.

Example 3.2
Consider a first-order continuous process

$$\dot{x}(t) = \alpha x(t) + \beta u(t) \tag{3.57a}$$

which is to be controlled. The EDS is (see Table 3.1)

$$x_{k+1} = e^{\alpha T_s} x_k + \frac{\beta}{\alpha}(e^{\alpha T_s} - 1)u_k \tag{3.57b}$$

To illustrate the dangers involved we proceed as follows. Let the sampling interval be very, very small so that $T_s \to 0$. As $T_s \to 0$, eqn. 3.57*b* shows that the state variable of the EDS becomes constant, in other words that the EDS is unreachable or uncontrollable. It might be concluded, incorrectly, that for small sampling intervals the process becomes difficult or impossible to control. Eqn. 3.57*a* shows that this is never so.

There can, however, be values of T_s which really do cause loss of reachability or observability when a continuous-time process is sampled. The following result (Kalman *et al.*, 1963; Gibson and Ha, 1980) holds for the time-invariant case.

Given the linear time-invariant process

$$\dot{x}(t) = Ax(t) + Bu(t)$$

Then necessary and sufficient conditions to ensure that the EDS is reachable are that

(i) $\text{rank}\,[B\ \Phi B\ \ldots\ \Phi^{n-1}B] = n$
(ii) if $\text{Re}\{\lambda_i(A)\} = 0$ for $i = 1, 2, \ldots, n$, then

$$\text{Im}\{\lambda_i(A)\} \neq \frac{2\pi k}{T_s}$$

Here, $\Phi = e^{AT_s}$, $k = \pm 1, \pm 2$, etc. and $n = \dim\{x(t)\}$.

Perhaps more insight can be gained into the effect of sampling by considering just the sufficient conditions for the preservation of reachability.

Given the controllable linear time-invariant process

$$\dot{x}(t) = Ax(t) + Bu(t)$$

Then the EDS is reachable if, when

$$\mathrm{Re}\{\lambda_i(A) - \lambda_j(A)\} = 0 \tag{3.58}$$

$$\mathrm{Im}(\lambda_i(A) - \lambda_j(A)\} \neq \frac{2\pi k}{T_s} \tag{3.59}$$

Thus, providing that the sampling interval T_s may be freely chosen it is always possible to satisfy eqn. 3.59. In that case a controllable continuous-time process remains controllable when it is incorporated into a sampled-data system. By duality, the same may be said of the observability (or reconstructibility) of a time-invariant process.

3.4.4 *Stabilisability and detectability*

It is worth recalling first how important (particularly in the discrete-time domain) these two properties are. Together, they turn out to be sufficient to guarantee that a controller or estimator applied to a process will function 'satisfactorily'. If a process is not stabilisable then, as the name suggests, no practical controller can ever make the process behaviour acceptable. If a process is not detectable, it is impossible to design a practical estimator to satisfactorily reconstruct the state of the system.

An introduction to stabilisability and detectability has already been given in Sections 2.3.5 and 2.3.7. Recall that stabilisability is a weaker condition than complete controllability (or reachability); detectability is weaker than complete reconstructibility (or observability). A stabilisable process is one in which the uncontrollable states are stable, i.e. decay harmlessly away. A detectable process is one where the unreconstructible states are stable. Stabilisability and detectability are dual concepts. Both together are sufficient to ensure the existence and the stability of the optimal controller and of the optimal estimator.

In this section we give two practical tests for checking either the stabilisability or the detectability of a time-invariant EDS. These tests are explicit in the system matrices. Then the discussion is broadened to include time-varying processes. Recall that in Chapter 2 only time-invariant processes were considered.

The following test (de Koning, 1982) is applicable to the time-invariant EDS.

The pair (Φ, Γ) is stabilisable if and only if

$$\lim_{i\to\infty} \|S_i\|^{1/i} < 1$$

where the $(n \times n)$ matrix S_i is found from

$$S_{i+1} = \Phi^T S_i \Phi - \Phi^T S_i \Gamma(\Gamma^T S_i \Gamma)^{-1}\Gamma^T S_i \Phi$$

and $S_0 = I$.

With appropriate modifications this test can be used to test detectability. Another possibility is the following (de Koning, 1983).

The pair (C, Φ) is detectable if and only if

$$\lim_{i\to\infty} \Phi^i e = 0$$

where $e = e_1 + e_2 + \ldots + e_k$ is the sum of the orthonormal eigenvectors corresponding to the zero eigenvalues of U, where

$$U = \sum_{i=0}^{n-1} \Phi^{Ti} C^T C \Phi^i$$

Again, this test could be adopted to test for stabilisability. Both this and the preceding test, it should be mentioned, are simplifications of tests that have been developed for systems with stochastic (multiplicative) parameters – under which the deterministic parameter systems of this book fall as a special case – and have therefore a wider significance.

Let us now turn to time-varying systems, in particular the general form of the EDS,

$$x_{k+1} = \Phi_k x_k + \Gamma_k u_k + w_k \tag{3.33a}$$

$$y_k = C_k x_k + v_k \tag{3.33b}$$

with which the reader is already familiar. We begin with the definition of *uniform detectability* (Anderson and Moore, 1981) for the system described by eqn. 3.33.

The pair (C_k, Φ_k) is uniformly detectable if there exist integers j, l with $j \geqslant k$, $l > k$ and constant scalars $0 \leqslant \alpha_0 < 1$, $\alpha_1 > 0$ such that whenever

$$\|\Phi(j, k)z\| \geqslant \alpha_0 \|z\| \tag{3.60}$$

for some z and integer k then

$$z^T \tilde{M}(k, l) z \geqslant \alpha_1 z^T z \tag{3.61}$$

where $\tilde{M}(k, l)$ is the discrete observability Gramian.

This definition can be made plausible as follows. The $(n \times n)$ observability Gramian $\tilde{M}(k, l)$, eqn. 3.56, is always non-negative definite. It will be positive definite only when all of the state trajectories are observable. Those unobservable trajectories must correspond to zero eigenvalues of $\tilde{M}(k, l)$; positive eigenvalues are associated with observable trajectories. For these latter it follows from the definition of eigenvalues that with z the corresponding eigenvector of $\tilde{M}(k, l)$:

$$\tilde{M}(k, l) z = \lambda(\tilde{M}) z \tag{3.62}$$

so that in the definition, eqn. 3.61 is picking out those observable trajectories of the process. Now if a trajectory is observable, we can, roughly speaking, do something with it – reconstruct it for example – even if the dynamics of that trajectory are unfavourable (even unstable). The unobservable trajectories must decay away quickly, i.e. be uniformly asymptotically stable. Eqn. 3.60 filters out all those state trajectories which do not decay fast enough. Thus the definition of uniform detectability is very similar to that presented already for time-invariant, continuous systems in Chapter 2.

The definition of *uniform stabilisability* is related to that of uniform detectability by a certain duality (Anderson and Moore, 1981). Here we would expect to find the discrete-time reachability Gramian, $\tilde{W}(k, l)$ of eqn. 3.55.

The pair (Φ_k, Γ_k) is uniformly stabilisable if there exist integers j, l with $j \geqslant k$, $l > k$ and constant scalars $0 \leqslant \alpha_0 < 1$, $\alpha_1 > 0$ such that whenever

$$\|\Phi(j, k)z\| \geqslant \alpha_0 \|z\| \qquad (3.63)$$

for some z and integer k then

$$z^T \tilde{W}(k, l)z \geqslant \alpha_1 z^T z$$

where $\tilde{W}(k, l)$ is the discrete reachability Gramian.

With uniform detectability and uniform stabilisability defined in this way it is possible to show (Anderson and Moore, 1981) that these together form sufficient conditions for the existence and the stability of both the optimal controller and the optimal estimator (also called the Kalman filter) for time-varying processes. We shall return to this point in later chapters.

References

ANDERSON, B. D. O. and MOORE, J. B. 1979: 'Optimal filtering' (Prentice-Hall)

ANDERSON, B. D. O. and MOORE, J. B. (1981): 'Detectability and stabilizability of time-varying discrete-time linear systems', *SIAM J. Control & Opt.*, **19**, 1, pp. 20–32

ASTRÖM, K. J. (1970): 'Introduction to stochastic control theory' (Academic Press)

ATHANS, M. (1971): 'The role and use of the stochastic linear-quadratic-gaussian problem in control system design', *IEEE Trans. Auto. Control*, **AC-16**, 6, pp. 529–552

DORATO, P. (1983): 'Theoretical developments in discrete-time control', *Automatica*, **19**, 4, pp. 395–400

EDGAR, T. F. (1982): 'Sampling and filtering for discrete-time systems', Am. Inst. Chem. Eng. Module A3.7., Process Control Modular Instruction Series

FRANKLIN, G. F. and POWELL, J. D. (1980): 'Digital control of dynamic systems' (Addison-Wesley)

GIBSON, J. A. and HA, T. T. (1980): 'Further to the preservation of controllability under sampling', *Int. J. Control*, **31**, 6, pp. 1013–1026

HALYO, N. and CAGLAYAN, A. K. (1976): 'A separation theorem for the stochastic sampled-data LQG problem', *Int. J. Control*, **23**, 2, pp. 237–244

JURY, E. I. and TSYPKIN, Ya. Z. (1971): 'On the theory of discrete systems', *Automatica*, **7**, pp. 89–107

KALMAN, R. E. and BERTRAM, J. E. (1960): 'Control system analysis and design via the "Second Method" of Lyapunov. Part II discrete-time systems', *Trans. ASME J. Basic Engng.*, **82**, pp. 394–400

KALMAN, R. E., HO, Y. C. and NARENDRA, K. S. (1963): 'Controllability of linear dynamical systems', *Contributions to Differential Equations*, **1**, 2, pp. 189–213

DE KONING, W. L. (1980): 'Equivalent discrete optimal control problem for randomly sampled digital control systems', *Int. J. Systems Sci.*, **11**, 7, pp. 841–850

DE KONING, W. L. (1982): 'Infinite horizon optimal control of linear discrete time systems with stochastic parameters', *Automatica*, **18**, 4, pp. 443–453

DE KONING, W. L. (1983): 'Detectability of linear discrete time systems with stochastic parameters', *Int. J. Control*, **38**, 5, pp. 1035–1047

KWAKERNAAK, H. and SIVAN, R. (1972): 'Linear optimal control systems' (John Wiley)

LANDAU, I. D. (1979): 'Adaptive control – the model reference adaptive approach', *Control and System Theory Series*, **8**, Marcel Dekker Inc.

LEVIS, A. H., SCHLUETER, R. A. and ATHANS, M. (1971): 'On the behaviour of optimal linear sampled-data regulators', *Int. J. Control*, **13**, pp. 343–361

MEDITCH, J. S. (1969): 'Stochastic optimal linear estimation and control' (McGraw-Hill)

MORI, T., FUKUMA, N. and KUWAHARA, M. (1982): 'Delay-independent stability criteria for discrete-delay systems', *IEEE Trans. Auto. Control*, **AC-27**, 4, pp. 964–966

OLIVER, B. M., PIERCE, J. R. and SHANNON, C. E. (1948): 'The philosophy of PCM', *Proc. IRE*, **36**, pp. 1324–1331

SMITH, C. L. (1972): 'Digital computer process control' (Intext Educational Publishers)

VERBRUGGEN, H. B., PEPERSTRAETE, J. A. and DEBRUIJN, H. P. (1975): 'Digital controllers and digital control algorithms', *Journal A*, **16**, 2, pp. 53–67

WIBERG, D. M. (1971): 'State space and linear systems' (McGraw-Hill)

Chapter 4

Parameter and state estimation

4.1 Introduction

So far it has been assumed that the process control engineer knows or can calculate all the parameters occurring in the process model discussed in the previous two chapters. Such an assumption is rarely justified in practice. Another, and related, problem is how to estimate the state of the process from observations or measurements made on that process. In this chapter we will address both of these problems and describe two, essentially similar, estimators which can be used for parameter and state estimation. By restricting the techniques of estimation to two, and frequently referring the reader for proofs and details to the extensive literature on this topic, the material in this chapter has been condensed to a minimal length. No historical perspective (Kalman, 1978; Sorenson, 1980) is provided.

From the outset we distinguish between two fundamentally different approaches to estimation. The first uses a deterministic process model and determines parameter values using realisation theory (Kalman and Ho, 1966) or reconstructs the state using observer theory (Luenberger, 1964). Deterministic realisation or observer theory has little relevance to the frequently changing and noisy environment of most estimation problems of process control.

The second approach to estimation assumes a stochastic process model, where there are noisy process observations or inaccurate measurements; we pursue this approach in the following five sections.

The estimation problem is solved when an acceptable approximation to an unknown set of parameters is found from observational data. The function of the observational data which is used to obtain the approximate value or *estimate* of the parameters set is called the *estimator*.

Before saying more about estimation, let us briefly consider the problems we wish to solve. The *parameter estimation problem* is to find the estimator that will produce the best estimates of all the process model parameters (and possibly the order of the model) from given observations or measurements. Knowledge of the state of the process, $x(t)$ or x_k, cannot be assumed. The *state estimation problem* is to find the estimator that will produce the best estimates of all the process model states from given measurements, assuming that all the model parameters are known. Adaptive state estimation, about which we will say nothing here, is concerned with establishing the state estimator when none of the model parameters is known.

Actually neither the parameter nor the state estimation problem is usually as severe as stated. In parameter estimation many of the parameters (and possibly the model order) are known or can be calculated; for example, parameters such as Henry's constant, the density of water etc. This knowledge must be used to maximum benefit. For example, by reducing the dimension of the problem the accuracy and speed of estimation can be increased.

Some terminology can usefully be introduced at this juncture. The state estimation can be classified as follows. If an estimate of the history of the state (i.e. the previous state trajectories) is required, state estimation is termed *smoothing*. *Prediction* is estimating the state in the future, while *filtering* is concerned with the present value of the state. The estimators are called smoothers, predictors and filters, respectively.

There are basically three approaches to obtain an estimate of some constant vector x from measurements y (Gelb, 1974).

> *Least squares estimation.* This is especially useful when so little information is available that there is no basis for assigning probability density functions (e.g. Gaussian, uniform etc.) to x or y.
> *Maximum likelihood estimation.* Some information, in the form of knowledge of the statistical properties of the measurement noise, is assumed to be available. The statistical properties of x, however, are unknown.
> *Bayesian estimation.* The statistical properties of both x and y are known. Of particular interest to us is the minimum variance estimator known as the Kalman filter.

It can be proved (Smith, 1965) that for Gaussian random variables all the above techniques produce identical estimates of x, provided that the assumptions are the same in each case. Since the Gaussian density function is encountered so frequently in theory and practice, this equality of estimates is extremely important.

Each estimator requires a model, however crude, relating measurements, unknowns to be estimated and noise. This model is the process (sometimes called the signal) model. Here, the process model is taken to be the finite-dimensional linear EDS introduced in the preceding chapter. Not only does this model represent practical situations but it furnishes an estimator which is tractible, and which can be implemented as an algorithm in a digital computer.

Pursuing this point, it is important to know what constitutes a good or desirable estimator. For ease of design and implementation the estimator should be *linear*, *finite-dimensional* and *discrete*, i.e. operate in the discrete time. In a *recursive form* the estimator requires less computer memory and can be used online. An operating criterion would be that the estimator is stable, producing estimates that are unbiased and converge to the true value in a reasonably short length of time. Moreover some degree of *robustness* is required, that is the estimator must maintain performance standards when assumed design parameters differ from those encountered in operation. These, and many other, benefits accrue from using an *optimal estimator* (Anderson and Moore, 1979), and in particular from using the estimator known as the Kalman (or Kalman–Bucy) filter.

Intuitively one feels that even such a desirable estimator might not be able to produce estimates in all situations. For example, it is difficult to see how an estimator could produce estimates of all the states of an unreconstructible or unobservable process. It turns out that if a practical estimator is required then only those processes which are uniformly stabilisable and detectable (Chapter 3) have states which can be estimated. If stabilisability and detectability are interpreted in the correct way then a similar statement holds for parameter estimation. The precise conditions will be given in this chapter. An allied problem is that of choosing the input signal to the process such that all the states or parameters can be estimated. This is the problem of choosing a persistently exciting input (Goodwin and Payne, 1977; Moore, 1983). In Section 4.5 we will show which conditions are necessary and sufficient for practical estimation.

It has already been stated that each estimator requires a process model. When the estimator has been designed and tested by performing simulations (Chapter 5) it will be applied to the actual process. Since the process model contains simplifying assumptions and perhaps inaccuracies it is important to be able to say whether the estimator is functioning correctly online. Filter performance will be discussed in the next two sections of this chapter.

In conclusion, a brief outline of the contents of this chapter is given. In Sections 4.2 and 4.3 two estimators, the (unsophisticated) least-squares estimator and the (sophisticated) Kalman filter are presented. These provide the means to solve our process control estimation problems. If possible the parameter estimation problem is solved using only steady-state measurements, Section 4.4. Often we must resort to measurements of process transients to provide the necessary information for parameter estimation: the Kalman filter is then applied to this problem in Section 4.5. Finally, state estimation with noisy measurements (if necessary artificially induced noise) is discussed in Section 4.6.

4.2 Least-squares estimation

The approach to estimation indicated in this section is the most basic possible; the least number of assumptions is made concerning the problem compatible with deriving a good estimator as discussed in the introduction to this chapter.

Here we analyse the linear least-squares problem, present the recursive least-squares estimator and examine its properties. Applications to process control of least-squares estimation are deferred until later in this chapter.

The linear least-squares problem may be stated as follows. The n-dimensional vector x contains n unknown constants which we wish to ascertain. In the absence of more information we can often do no better than to assume an unknown quantity is constant. There are m sensors making periodic measurements so that at any instant of time kT_s ($k = 0, 1, 2, \ldots$) there are m observations, collected together in the m-dimensional vector y_k. The first measurements available are therefore to be found in y_0. It is assumed that every sensor output, $y_k^{(i)}$, is linearly related to the unknowns $x^{(1)}, x^{(2)}, \ldots, x^{(n)}$ (elements of the vector x) by

$$y_k^{(i)} = c_{1k}^{(i)}x^{(1)} + c_{2k}^{(i)}x^{(2)} + \ldots + c_{nk}^{(i)}x^{(n)} + v_k^{(i)} \tag{4.1a}$$

or

$$y_k = C_k x + v_k \tag{4.1b}$$

where v_k is an $(m \times 1)$ vector of unknown errors occurring in the measurement of x. C_k is a $(m \times n)$ matrix of known elements. In general $n > m$, i.e. there are more unknowns than sensors. Now if we take measurements at M instants in time ($k = 0, 1, 2, \ldots, M - 1$) then at a certain moment $Mm > n$ and we ask whether an estimator exists whose estimate of x minimises the effect of the measurement errors, v_k. The M equations having the form of eqn. 4.1b are first accumulated in an $(Mm \times 1)$ vector z:

$$\begin{bmatrix} y_0 \\ y_1 \\ \cdot \\ \cdot \\ \cdot \\ y_{M-1} \end{bmatrix} = \begin{bmatrix} C_0 \\ C_1 \\ \cdot \\ \cdot \\ \cdot \\ C_{M-1} \end{bmatrix} x + \begin{bmatrix} v_0 \\ v_1 \\ \cdot \\ \cdot \\ \cdot \\ v_{M-1} \end{bmatrix} \tag{4.2a}$$

i.e.

$$z = Hx + v \tag{4.2b}$$

The least-squares estimator can be stated as follows (Junkins, 1978).

Given the process model

$$z = Hx + v \tag{4.2b}$$

and the cost criterion

$$J(a) = (z - Ha)^{\mathrm{T}} R (z - Ha) \tag{4.3}$$

where R is a $(Mm \times Mm)$ positive definite, M block-diagonal, weighting matrix. Then if

$$\text{rank}\,(H) = n \tag{4.4}$$

the estimate a of x which minimises eqn. 4.3 exists, and is given by the least-squares estimator:

$$\hat{x} = (H^{\mathrm{T}}RH)^{-1}H^{\mathrm{T}}Rz \tag{4.5}$$

and $\hat{x}$ is the least-squares estimate of x.

Loosely speaking the process model and the cost criterion (sometimes called the performance criterion or index) together define the optimisation problem solved to provide the least-squares estimate. Since not all measurement errors need be of equal importance we have allowed the errors to be weighted by including the matrix R is the cost criterion. Its structure is due to the assumed independence of each set of m measurement errors from preceding or following errors. The minimum value of the cost criterion is, from eqns. 4.3 and 4.5,

$$J_{\min} = z^{\mathrm{T}}[R - RH(H^{\mathrm{T}}RH)^{-1}H^{\mathrm{T}}R]z \tag{4.6}$$

We see that if, for example, $R = \alpha^2 I$, α a nonzero constant, then the estimate of x remains unchanged when R is scaled-up, although the minimum cost increases proportionally.

The rank condition, eqn. 4.4, ensures the existence of the inverse quantity appearing in the estimator. This is not to say that calculation of this quantity is without numerical difficulties, but a number of techniques exist to alleviate this problem (see, for example, Lawson and Hanson, 1974; Sorenson, 1980). Note that if $C_0 = C_1 = \ldots = C_{M-1}$ and are time-invariant then the rank condition is only satisfied if $m = n$, i.e. there must be as many measurement sensors as unknown constants.

The least-squares estimator is suitable for batch processing of data and may be useful for some offline applications. For online estimation, where measurements are continually becoming available and an estimate must be found in real-time, the estimator just presented is cumbersome and inefficient, since no old measurements are discarded once they have been processed. The *recursive* (or sequential) *least-squares estimator* (*RLS*) solve this problem (Hastings-James and Sage, 1969; Junkins, 1978).

Suppose we take measurements at two instants of time $k = 0, 1$. Some useful notation is now introduced. Let $\hat{x}_{i/j}$ be the estimate of x made at time iT_s using measurements $y_0, y_1, \ldots, y_j$. With this notation and eqn. 4.5 the first two estimates of the least-squares estimator could be written

$$\hat{x}_{0/0} = (C_0^{\mathrm{T}}R_0C_0)^{-1}C_0^{\mathrm{T}}R_0y_0 \tag{4.7}$$

$$\hat{x}_{1/1} = (C_0^{\mathrm{T}}R_0C_0 + C_1^{\mathrm{T}}R_1C_1)^{-1}(C_0^{\mathrm{T}}R_0y_0 + C_1^{\mathrm{T}}R_1y_1) \tag{4.8}$$

provided that $(C_0^{\mathrm{T}}R_0C_0)$ is nonsingular. Here, R_i is the ith block diagonal element of R. Define a new $(n \times n)$ matrix Σ_k by

$$\Sigma_k \triangleq (C_0^T R_0 C_0 + C_1^T R_1 C_1 + \ldots + C_k^T R_k C_k)^{-1} \quad (4.9)$$

then by generalising eqns. 4.7 and 4.8 for the k and $k + 1$ step we find that

$$\hat{x}_{k+1/k+1} = \hat{x}_{k/k} + \Sigma_{k+1} C_{k+1}^T R_{k+1}(y_{k+1} - C_{k+1}\hat{x}_{k/k}) \quad (4.10)$$

This recursive expression is the RLS. It must be correctly initialised and provided with a recursive algorithm for computing Σ_{k+1} from Σ_k. The latter is obtained by applying the matrix inversion lemma (Chapter 1) to eqn. 4.9. The result is summarised as follows.

Let every available set of measurement data, y_k, be known to be related to the unknown constant vector x by

$$y_k = C_k x + v_k \quad (4.1b)$$

Then provided

$$\text{rank}\ (C_0) = n \quad (4.11)$$

the estimate a_k of x which minimises the cost criterion

$$J_k(a_k) = \sum_{i=0}^{k} (y_i - C_i a_i)^T R_i (y_i - C_i a_i) \quad (4.3)$$

is given by the *recursive least-squares estimator* (RLS)

$$\hat{x}_{k+1/k+1} = \hat{x}_{k/k} + K_{k+1}(y_{k+1} - C_{k+1}\hat{x}_{k/k}) \quad (4.12)$$

$$K_k = \Sigma_k C_k^T R_k \quad (4.13)$$

$$\Sigma_{k+1} = \Sigma_k - \Sigma_k C_{k+1}^T (C_{k+1}\Sigma_k C_{k+1}^T + R_{k+1}^{-1})^{-1} C_{k+1}\Sigma_k \quad (4.14)$$

Here, R_k is the positive definite kth block of R, eqn. 4.3. The RLS is initialised with

$$\hat{x}_{0/0} = (C_0^T R_0 C_0)^{-1} C_0^T R_0 y_0 \quad (4.15)$$

$$\Sigma_0 = (C_0^T R_0 C_0)^{-1} \quad (4.16)$$

$\hat{x}_{k/k}$ is the recursive least-squares estimate of x.

Each new estimate $\hat{x}_{k+1/k+1}$ is found from the previous estimate $\hat{x}_{k/k}$ by adding a correction term, which is the *gain matrix* K_{k+1} times the residue

$$y_{k+1} - C_{k+1}\hat{x}_{k/k}$$

The RLS is seen to be a linear, time-varying, deterministic system. Note that R_k is always positive definite since R being positive definite implies that the determinant of R must be positive. Moreover, since we must assume that C_0 has rank n (hence $m \geqslant n$) to permit inversion of Σ_0 in eqns. 4.15 and 4.16, it follows that Σ_k is a (symmetric) positive definite matrix for all finite k. This can be seen from eqn. 4.9 and the fact that the sum of a positive and a non-negative definite matrix is itself positive definite (Chapter 1).

It must be emphasised that the estimator derived is actually a filter (see the previous section). Suppose we wished to predict the value of x one step further in time, i.e. find $\hat{x}_{k+1/k}$. Recall that x has been assumed to be constant, so clearly

$$\hat{x}_{k+1/k} = \hat{x}_{k/k} \tag{4.17}$$

The (one-step) predictor is found from eqns. 4.12 and 4.17 to be

$$\hat{x}_{k+1/k} = \hat{x}_{k/k-1} + K_k(y_k - C_k\hat{x}_{k/k-1}) \tag{4.18}$$

This form of the RLS is important because it is so similar to the form of the Kalman filter (Section 4.3).

Example 4.1
To illustrate the use of the batch and recursive forms of the estimator we consider the problem of estimating the *sample mean*. There are assumed to be m measurements, each related to the scalar unknown $x^{(1)}$ according to (see eqn. 4.1a)

$$y^{(i)} = x^{(1)} + v^{(i)} \tag{4.19a}$$

so that if $H = [1\ 1\ 1\ \ldots\ 1]^T$ is an $(m \times 1)$ matrix

$$y = Hx^{(1)} + v \tag{4.19b}$$

Taking $R = \alpha^2 I$ the least-squares estimator, eqn. 4.5 may be found to be given by

$$\hat{x} = \frac{1}{m}\sum_{i=0}^{m=1} y^{(i)} \tag{4.19c}$$

which is the expression usually used to calculate the sample mean. Consider next the processing of each measurement recursively. Eqn. 4.19a becomes the RLS process model (eqn. 4.1b) so that the observation matrix C_k is a scalar equal to one. From eqn. 4.14, with R also a (nonzero) scalar,

$$\begin{aligned}\Sigma_{k+1} &= \Sigma_k - \Sigma_k^2/(\Sigma_k + R^{-1}) \\ &= \Sigma_k R^{-1}/(\Sigma_k + R^{-1})\end{aligned} \tag{4.19d}$$

The RLS gain matrix is therefore given by

$$K_{k+1} = \Sigma_k/(\Sigma_k + R^{-1}) \tag{4.19e}$$

initialised with $\Sigma_0 = R^{-1}$. It is easy to show that $K_{k+1} = 1/(k + 1)$ so that the RLS becomes

$$\hat{x}_{k+1/k+1} = \hat{x}_{k/k} + (y^{(k)} - \hat{x}_{k/k})/(k + 1) \tag{4.19f}$$

with

$$\hat{x}_{0/0} = y^{(0)} \tag{4.19g}$$

After m measurements have been made, the two forms of the least-squares estimator provide the same estimate $\hat{x}$ of x.

We now investigate some properties of the RLS; to do this we assume that more and more measurements are taken, so that $k \rightarrow \infty$. We use the predictor form of the RLS, eqn. 4.18, since this aids our insight. Denote the estimator error by $\tilde{x}_k$, where

$$\tilde{x}_k \triangleq x - \hat{x}_{k/k+1} \tag{4.20}$$

Subtracting both sides of eqn. 4.18 from x and using eqn. 4.1b gives

$$\tilde{x}_{k+1} = (I - K_k C_k)\tilde{x}_k - K_k v_k \tag{4.21}$$

and the corresponding unforced system is

$$\tilde{x}_{k+1} = (I - K_k C_k)\tilde{x}_k \tag{4.22}$$

Clearly, if eqn. 4.22 is uniformly asymptotically stable then as k increases, $\tilde{x}_k \rightarrow 0$ so that $\hat{x}_k$ will converge to x. Note that the unforced system is time-varying even if C_k is time-invariant. Now by using the matrix inversion lemma (Chapter 1), an alternative form can be found for the gain matrix K_k. From eqns. 4.13 and 4.14,

$$\begin{aligned} K_k &= (\Sigma_{k-1} - \Sigma_{k-1}C_k^{\mathrm{T}}[C_k\Sigma_{k-1}C_k^{\mathrm{T}} + R_k^{-1}]^{-1}C_k\Sigma_{k-1})C_k^{\mathrm{T}}R_k \\ &= (\Sigma_{k-1}^{-1} + C_k^{\mathrm{T}}R_kC_k)^{-1}C_k^{\mathrm{T}}R_k \\ &= \Sigma_{k-1}C_k^{\mathrm{T}}(C_k\Sigma_kC_k^{\mathrm{T}} + R_k^{-1})^{-1} \end{aligned} \tag{4.23}$$

Hence eqn. 4.14 can be written

$$\Sigma_k = (I - K_k C_k)\Sigma_{k-1} \tag{4.24}$$

Comparing eqns. 4.24 and 4.22 we see that, roughly speaking, the dynamics of Σ_k and $\tilde{x}_k$ are the same. We shall exploit this similarity shortly. First, recall our definition of Σ_k, eqn. 4.9. If we impose the restriction that

$$C_k^{\mathrm{T}}R_kC_k \geqslant \alpha I > 0 \tag{4.25}$$

for all k, then as $k \rightarrow \infty$, $\Sigma_k \rightarrow 0$. This is apparent from eqn. 4.9 where we see that

$$\Sigma_k^{-1} = (C_k^{\mathrm{T}}R_0C_0 + C_1^{\mathrm{T}}R_1C_1 + \ldots + C_k^{\mathrm{T}}R_kC_k) \geqslant (k+1)\alpha I$$

so that

$$\Sigma_k \leqslant \frac{1}{(k+1)\alpha} I \tag{4.26}$$

Hence with the restriction that eqn. 4.25 is satisfied, Σ_k converges to zero. Now from the repeated application of eqn. 4.24 we see that

$$\Sigma_k = A_k A_{k-1} \ldots A_0 \Sigma_0 \tag{4.27}$$

where $A_k \triangleq I - K_k C_k$. Since Σ_0 can take on any value and yet $\Sigma_k \to 0$ as $k \to \infty$, we conclude that as $k \to \infty$

$$A_k A_{k-1} \ldots A_0 \to 0 \tag{4.28}$$

However, the propagation of the estimator error, given by eqn. 4.22, can also be written as

$$\tilde{x}_{k+1} = A_k A_{k-1} \ldots A_0 \tilde{x}_0 \tag{4.29}$$

which in view of eqn. 4.28 shows that as $k \to \infty$, $\tilde{x}_k \to 0$. This result is summarised in the following:

If $C_k^{\mathrm{T}} R_k C_k \geqslant \alpha I > 0$ for all k then the RLS is stable, that is to say that the estimate of $\tilde{x}_k$ of x converges to x as $k \to \infty$.

The nonconvergence of the RLS to the true x under certain circumstances has been cited incorrectly by some authors as an inherent weakness of the RLS. In fact a sufficient condition which guarantees the good functioning of the RLS has just been presented.

The test says nothing, of course, about the rate of convergence of $\hat{x}_k$ to x, although it is worth noting that the gain matrix and Σ_k can be computed in advance of any measurements becoming available. Moreover, some check is required as to whether the RLS is performing correctly. If $C_k^{\mathrm{T}} R_k C_k \geqslant \alpha I > 0$ and $\hat{x}_k$ converges to x as $k \to \infty$, then $\Sigma_k \to 0$ and, from eqn. 4.13 we see that $K_k \to 0$. This means that the RLS becomes 'saturated', that is, it makes little or no use of information contained in new measurements. To counteract this, but more importantly to speed the rate of convergence, a so-called *forgetting factor* may be incorporated into the RLS (see, for example, Eykhoff, 1974). A similar technique, called exponential data weighting (Anderson and Moore, 1979), has been proposed for the Kalman filter.

Of course there are applications of the RLS where there is a possibility that the basic assumption underlying the model may be violated, e.g. x not a constant. However, if the dynamic variation of x can be described by a known linear model it is still possible to apply the RLS theory to achieve an optimal estimate (see, for example, Junkins, 1978). That apart, it is difficult to derive any checks on the RLS performance without making more assumptions concerning the measurement errors v_k. This, in effect, is what we do in the next section.

Finally, we remark that an extension of this estimation technique to non-linear process models is possible (Bard, 1974; Junkins, 1978). However, the estimator is not then optimal, in the sense of a cost criterion being minimised, nor are the convergence conditions straightforward. The appeal of the linear least-squares estimator is that the assumptions made concerning the process model and the cost criterion are an absolute minimum. Adding extra structure to the process model detracts from this basic simplicity.

4.3 Minimum variance estimation – the Kalman filter

The process model of the last section was extremely unsophisticated: it was a static model and it was deterministic. Moreover, no prior information concerning measurement errors or the value of x was assumed to be available. Here we will be concerned with a dynamic process model with additive stochastic noise for which we are given *a priori* information; in fact the process model is the familiar EDS of Chapter 3. Instead of estimating unknown constants, we now estimate the model state, x_k. In this section we introduce the linear minimum variance problem, present the recursive minimum variance estimator, or Kalman filter (KF) as it is called, and examine its properties. Applications of the KF to process control are examined in Sections 4.5 and 4.6.

Suppose we have the same process model as in the previous section but that we allow $\{v_k\}$, the measurement error, to be a random, zero-mean vector. The cost criterion to be minimised is not that of the previous section but the expected value of the norm of the estimation error, i.e. $\|x - a\|^2$. Then the Gauss–Markov theorem, proved below, provides the optimal linear estimator.

Given the process model

$$z = Hx + v \tag{4.30}$$

where $E\{v\} = 0$ and $E\{vv^T\} = V$, where $V > 0$. The cost criterion is

$$J(a) = E\{(x - a)^T Q(x - a)\} \tag{4.31}$$

where Q is a $(n \times n)$ positive definite weighting matrix. Then if

$$\text{rank}\ (H) = n \tag{4.32}$$

the optimal linear estimate $a = \hat{x}$ of x exists and is given by the *linear minimum variance estimator*:

$$\hat{x} = (H^T V^{-1} H)^{-1} H^T V^{-1} z \tag{4.33}$$

where $\hat{x}$ is the linear minimum variance estimate of x.

Proof:

Assume the estimator has a linear form,

$$a = Az \tag{4.34}$$

where A is an $(n \times Mm)$ matrix. Then

$$E\{a\} = AE\{z\} = AHx$$

since $E\{v\} = 0$. If the estimate is to be unbiased we must have $E\{a\} = x$ for any x so that

$$AH = I \tag{4.35}$$

Now

$$\begin{aligned} J(a) &= E\{(x - Az)^{\mathrm{T}}Q(x - Az)\} \\ &= E\{(Hx - z)^{\mathrm{T}}A^{\mathrm{T}}QA(Hx - z)\} \\ &= E\{v^{\mathrm{T}}A^{\mathrm{T}}QAv\} \\ &= E\{\mathrm{tr}(A^{\mathrm{T}}QAvv^{\mathrm{T}})\} \\ &= \mathrm{tr}(A^{\mathrm{T}}QAV) \end{aligned} \tag{4.36}$$

We wish to determine the matrix A which minimises eqn. 4.36 subject to 4.35. Construct the Lagrangian

$$L = \mathrm{tr}(A^{\mathrm{T}}QAV) + \lambda^{\mathrm{T}}(AH - I)\lambda \tag{4.37}$$

where λ is a $(n \times 1)$ Lagrangian multiplier. Clearly, minimising eqn. 4.37 with respect to A would achieve our purpose. A stationary point of the above equation satisfies

$$\frac{\partial L}{\partial A} = 2QAV + \lambda\lambda^{\mathrm{T}}H^{\mathrm{T}} = 0 \tag{4.38}$$

so that using eqn. 4.35

$$2Q + \lambda\lambda^{\mathrm{T}}H^{\mathrm{T}}V^{-1}H = 0$$

In view of the rank condition on H,

$$\lambda\lambda^{\mathrm{T}} = -2Q(H^{\mathrm{T}}V^{-1}H)^{-1} \tag{4.39}$$

which when substituted back into eqn. 4.38 gives

$$A = (H^{\mathrm{T}}V^{-1}H)^{-1}H^{\mathrm{T}}V^{-1}$$

It is not difficult to verify that this solution is indeed a minimum of eqn. 4.36, which proves the theorem.

The proof just given provides us with the optimum *linear* estimator. Therefore, there could exist a nonlinear estimator capable of making better estimates of x. However, in one very important practical situation the estimator just given is *the* optimal estimator (Anderson and Moore, 1979).

If, in addition to the assumptions given concerning the random vector v of the process model, v is *normally distributed* (Gaussian) then eqn. 4.33 is the optimal estimator.

The linear minimum variance estimator is remarkably similar to the linear least-squares estimator of Section 4.2. Notice that although both cost criteria are deterministic scalar quantities, they are fundamentally different. Moreover, the value of the weighting matrix Q of eqn. 4.31 is immaterial to the estimator, and the weighting matrix of the least-squares estimator may be identified with the covariance matrix of the process model noise, v. The minimum cost associated with the linear minimum variance estimator is found by substituting eqn. 4.33 into 4.31:

$$\begin{aligned} J_{\min} &= E\{v^{\mathrm{T}}V^{-1}H(H^{\mathrm{T}}V^{-1}H)^{-1}Q(H^{\mathrm{T}}V^{-1}H)^{-1}H^{\mathrm{T}}V^{-1}v\} \\ &= E\{\mathrm{tr}[V^{-1}H(H^{\mathrm{T}}VH)^{-1}Q(H^{\mathrm{T}}V^{-1}H)^{-1}H^{\mathrm{T}}V^{-1}vv^{\mathrm{T}}\} \\ &= \mathrm{tr}[V^{-1}H(H^{\mathrm{T}}VH)^{-1}Q(H^{\mathrm{T}}V^{-1}H)^{-1}H^{\mathrm{T}}] \\ &= \mathrm{tr}[Q(H^{\mathrm{T}}V^{-1}H)^{-1}] \end{aligned} \tag{4.40}$$

and this has the obvious interpretation that the more noise there is that corrupts the process model the more difficult it is to estimate x.

Since our intention is to shortly consider the more general dynamic process model with *a priori* statistical information we will not delve deeply into other interesting aspects of the optimal minimum variance estimator just presented. It is necessary, however, to recall the remark made in the preceding section concerning the conditional nature of the estimate $\hat{x}$. Also, from eqns. 4.33 and 4.30 we see that the estimator is a function of a random variable, v, and is therefore itself a random variable, with mean x and covariance $(H^{\mathrm{T}}V^{-1}H)^{-1}$.

Just as the RLS was developed in Section 4.2 it is possible to develop a recursive form of the minimum variance estimator. If at the same time the process model is generalised to a discrete-time stochastic non-stationary dynamic model then the estimator which results is called the *Kalman* (*or Kalman–Bucy*) *filter* (*KF*). A derivation of the KF is not provided; we give here only a concise statement of the chief results (Kalman, 1960; Anderson and Moore, 1979).

Given the process model

$$x_{k+1} = \Phi_k x_k + \Gamma_k u_k + w_k \tag{4.41}$$

$$y_k = C_k x_k + v_k \tag{4.42}$$

$$E\{x_0\} = \bar{x}_0 \tag{4.43}$$

$$E\{(x_0 - \bar{x}_0)(x_0 - \bar{x}_0)^{\mathrm{T}}\} = \Sigma_0 \tag{4.44}$$

$$E\{w_k\} = 0 \tag{4.45}$$

$$E\{v_k\} = 0 \tag{4.46}$$

$$E\{w_k w_l^{\mathrm{T}}\} = W_k \delta_{kl}, \quad W_k \geqslant 0 \tag{4.47}$$

$$E\{w_k v_l^{\mathrm{T}}\} = S_k \delta_{kl} \tag{4.48}$$

$$E\{v_k v_l^{\mathrm{T}}\} = V_k \delta_{kl}, \quad V_k \geqslant 0 \tag{4.49}$$

where u_k is a deterministic variable (possibly dependent on past controls and measurements). x_0 is not correlated with w_k or with v_k, where $\{w_k\}$ is the (white) process noise and $\{v_k\}$ the (white) measurement noise, possibly correlated to each other through eqn. 4.48. Then the *linear minimum variance estimator* (or

KF) is given by:

$$\hat{x}_{k+1/k} = \Phi_k \hat{x}_{k/k-1} + \Gamma_k u_k + K_k(y_k - C_k \hat{x}_{k/k-1}) \quad (4.50)$$

$$\hat{x}_{0/-1} = \bar{x}_0 \quad (4.51)$$

$$K_k = (\Phi_k \Sigma_{k/k-1} C_k^{\mathrm{T}} + S_k)(C_k \Sigma_{k/k-1} C_k^{\mathrm{T}} + V_k)^{-1} \quad (4.52)$$

$$\Sigma_{k+1/k} = \Phi_k \Sigma_{k/k-1} \Phi_k^{\mathrm{T}} - (\Phi_k \Sigma_{k/k-1} C_k^{\mathrm{T}} + S_k) \times (C_k \Sigma_{k/k-1} C_k^{\mathrm{T}} + V_k)^{-1} (\Phi_k \Sigma_{k/k-1} C_k^{\mathrm{T}} + S_k)^{\mathrm{T}} + W_k \quad (4.53)$$

$$\Sigma_{0/-1} = \Sigma_0 \quad (4.54)$$

and $\hat{x}_{k/k-1}$ is the linear minimum variance of x_k at time kT_s given the measurements $y_0, y_1, \ldots, y_{k-1}$.

The process model is the now familiar EDS. The only restriction we need place on the process parameters is that they are available for computation in the estimator at time kT_s; they may therefore comprise past measurements, certain control signals etc. Time-varying noise covariances W_k, S_k and V_k are permitted – in which case the model is said to be *nonstationary* – but W_k and V_k are of course non-negative definite for all k. If the matrix $C_k \Sigma_{k/k-1} C_k^{\mathrm{T}} + V_k$ is singular so that the inverse occurring in eqns. 4.45 and 4.46 does not exist, there is said to be a *singular estimation problem*. Note that if $V_k > 0$ for all k the inverses will always exist. One remedy is to replace non-existent inverses with their pseudo-inverses (Clements and Anderson, 1978). Another approach, whereby the state dimension is reduced, is presented in Section 4.6.

There is a striking resemblance between the RLS of the last section and the KF. In particular:

The KF with $\Phi_k = I, \Gamma_k = W_k = S_k = 0, V_k = R_k^{-1}$ and subject to the initial conditions $\bar{x}_0 = 0$, $\Sigma_0^{-1} = 0$ is the same as the predictor form of the RLS, eqn. 4.18.

We will return to $\Sigma_0^{-1} = 0$ shortly.

The quantity K_k of eqn. 4.52 is called the *Kalman filter gain*, and $\Sigma_{k/k-1}$ the *error covariance*. This latter name can be made plausible by computing the covariance of the error $\tilde{x}_k$ (see, eqn. 4.20):

$$\tilde{x}_k \triangleq x_k - \hat{x}_{k/k-1} \quad (4.55)$$

From eqns. 4.50 and 4.41,

$$\tilde{x}_{k+1} = \Phi_k \tilde{x}_k - K_k(y_k - C_k \hat{x}_{k/k-1}) + w_k \quad (4.56)$$

and by eqn. 4.42

$$\tilde{x}_{k+1} = (\Phi_k - K_k C_k)\tilde{x}_k + w_k - K_k v_k \quad (4.57)$$

Hence

$$E\{\tilde{x}_{k+1}\tilde{x}_{k+1}^{\mathrm{T}}\} = (\Phi_k - K_kC_k)E\{\tilde{x}_k\tilde{x}_k^{\mathrm{T}}\}(\Phi_k - K_kC_k)^{\mathrm{T}} \times [I - K_k]\begin{bmatrix} W_k & S_k \\ S_k^{\mathrm{T}} & V_k \end{bmatrix}\begin{bmatrix} I \\ -K_k \end{bmatrix} \tag{4.58}$$

since $\tilde{x}_k$ and $[w_k^{\mathrm{T}} v_k^{\mathrm{T}}]^{\mathrm{T}}$ can be shown to be uncorrelated and have zero mean. After some manipulation, eqn. 4.58 can be written in the same form as eqn. 4.53, so that

$$\Sigma_{k/k-1} = E\{\tilde{x}_k\tilde{x}_k^{\mathrm{T}}\} \tag{4.59}$$

The error covariance, $\Sigma_{k/k-1}$, is non-negative definite, and can be precomputed (see the preceding section). Eqn. 4.53, called the *discrete-time Riccati equation*, is one of the most important in all control engineering, and has been extensively studied.

If there is no prior knowledge concerning the value initially taken by x_k, it is necessary to set $\Sigma_{0/-1}^{-1} = \Sigma_0^{-1} = 0$. A special form of the KF called the Information Filter (Anderson and Moore, 1979) can handle this situation. Of course, if the KF is in use long enough the effect of the initial transients, and thus the value chosen for $\bar{x}_0$ and Σ_0, become of little importance.

The KF is an *unbiased estimator*. Taking expectations of both sides of eqn. 4.57 gives

$$E\{\tilde{x}_{k+1}\} = (\Phi_k - K_kC_k)E\{\tilde{x}_k\} \tag{4.60}$$

Using eqn. 4.51 we have that

$$E\{\tilde{x}_0\} = E\{x_0 - \hat{x}_{0/-1}\} = E\{x_0 - \bar{x}_0\} = 0 \tag{4.61}$$

so that from eqn. 4.60

$$E\{\hat{x}_{k/k-1}\} = E\{x_k\}, \quad \text{for all } k \tag{4.62}$$

which is what we mean by an unbiased estimate.

In one important practical situation the KF is not only the best linear estimator, but the best estimator.

If, in addition to the assumptions given concerning x_0, $\{w_k\}$ and $\{v_k\}$ of the process model, it is assumed that they are jointly normally distributed (Gaussian) then eqn. 4.50 is the optimal estimator.

$\hat{x}_{k/k-1}$ is the minimum variance prediction of x.

We now turn to the performance of the KF. It is necessary, at least mathematically, to distinguish between the KF behaviour for a finite number of sampling instants, and its behaviour as $k \to \infty$. The latter enables the stability of the KF to be introduced. Then if the KF exists and moreover is stable, the behaviour of the KF may be split into a transient and a steady-state phase.

First we must clarify the concept of stability as applied to the KF. The KF is a time-varying, stochastic, linear, dynamic system which may be written (from eqn. 4.50) as

$$\hat{x}_{k+1/k} = (\Phi_k - K_k C_k)\hat{x}_{k/k-1} + \Gamma_k u_k + K_k y_k$$

The unforced KF is found by dropping the last two terms

$$\hat{x}_{k+1/k} = (\Phi_k - K_k C_k)\hat{x}_{k/k-1}$$

and it is this deterministic system (since $\hat{x}_{0/-1}$ is deterministic so are all $\hat{x}_{k+1/k}$) which is studied in connection with the stability of the KF. This is the same equation (but with a different initial condition) as must be used to study the stability of the error, $\tilde{x}_k$, propagation (see eqn. 4.57). Note that the forcing terms, $\Gamma_k u_k$ and $K_k y_k$ of the KF result not only in $\hat{x}_{k+1/k}$ being a stochastic variable but also account for $\hat{x}_{k+1/k}$ attaining (possibly) nonzero equilibrium or steady states.

The conditions which ensure that the KF exists and moreover is stable are now given (Anderson and Moore, 1981):

Let the process model, eqns. 4.41 to 4.49, with Φ_k, Γ_k, C_k, W_k, S_k and V_k all uniformly bounded, be such that $V_k \geqslant \alpha I > 0$. Let $W_k^{1/2}$ be a square-root of W_k (see Chapter 1). Then

(a) If $(C_k^{\mathrm{T}}, \Phi_k)$ is uniformly detectable, then $\Sigma_{k/k-1}$ is bounded for all k.

(b) If, in addition, $(\Phi_k, W_k^{1/2})$ is uniformly stabilisable, the KF is uniformly asymptotically stable.

Stabilisability and detectability have already been discussed in Section 3.4.4. It should be obvious that if the process model itself is uniformly asymptotically stable (with bounded parameters and $V_k \geqslant \alpha I > 0$) then both the above conditions are fulfilled.

Frequently the process model parameters are time invariant and the noise sources stationary. Because of the time dependence of the KF gain K_k, the KF is nevertheless still a time-varying dynamic system. If we now ensure that the KF exists and is stable then we can obtain some simplification as the following result (see, for example, Anderson and Moore, 1979) indicates.

Let the process model be taken as eqns. 4.41 to 4.49 with all the process model parameters and noise statistics time-invariant. Let $V > 0$, $(\Phi, W^{1/2})$ be stabilisable and (C^{T}, Φ) be detectable. Then as $k \to \infty$, $\Sigma_{k/k-1}$ converges to the unique, non-negative definite solution, Σ, of the *Algebraic Riccati Equation*:

$$\Sigma = \Phi\Sigma\Phi^{\mathrm{T}} - (\Phi\Sigma C^{\mathrm{T}} + S)(C\Sigma C^{\mathrm{T}} + V)^{-1}(\Phi\Sigma C^{\mathrm{T}} + S)^{\mathrm{T}} + W \tag{4.63}$$

The corresponding filter gain matrix, the *steady-state KF gain* is

$$K = (\Phi\Sigma C^{\mathrm{T}} + S)(C\Sigma C^{\mathrm{T}} + V)^{-1} \tag{4.64}$$

The properties and solution methods for the Algebraic Riccati Equation have been extensively studied. Since it is a nonlinear equation, other, but not non-negative definite, solutions to it may exist. In general (choose for example a stable model with $\Phi = S = 0$) the unique non-negative solution Σ is non-zero, so that although the estimate is unbiased (eqn. 4.62) the mean of the error covariance (eqn. 4.59) is generally non-zero. However, the steady-state KF gain may well be zero, in which case the KF is said to be saturated since new measurements do not affect the estimate. Various techniques (see Anderson and Moore, 1979) have been proposed to overcome saturation.

The above result clarifies our ideas, at least for time-invariant process models, concerning the performance of the KF. Providing $V > 0$ and the stabilisability and detectability constraints are met, there is indeed a transient phase of behaviour leading to a steady state. Since the steady-state filter gain is precomputable it is possible to construct, using eqns. 4.50 and 4.64, a very simple estimator (the algebraic or *steady-state KF*). Its structure is similar to that of an observer (Luenberger, 1964) and it will prove useful in many applications. The speed of response of the steady-state KF is found by evaluating the eigenvalues of the matrix:

$$\Phi - (\Phi\Sigma C^{\mathrm{T}} + S)(C\Sigma C^{\mathrm{T}} + V)^{-1}C$$

If the filter performs sluggishly, exponential data weighting (see, for example, Anderson and Moore, 1979) or an artificial increase in W can lead to better filter performance.

Another topic of paramount importance is the correct or otherwise functioning of the KF *in situ*. Recall that the KF was designed using process model data and not data from the process itself (although that can also be done as explained in Section 4.6). We introduce the *innovations* (or perhaps more correctly the innovations sequence) defined as z_k:

$$z_k \triangleq y_k - C_k\hat{x}_{k/k-1} \tag{4.65}$$

The innovations represent the new information carried in y_k but not in preceding measurements. It can be proved (Anderson and Moore, 1979) that the innovations have the following properties.

Define the innovations of the KF by eqn. 4.65. Then

1 $$E\{z_k\} = 0 \tag{4.66}$$

2 $$E\{z_k z_l^{\mathrm{T}}\} = (C_k\Sigma_{k/k-1}C_k^{\mathrm{T}} + V_k)\delta_{kl} \tag{4.67}$$

The performance of the KF can be monitored if the sample-mean, sample-covariance and whiteness of the innovations are recorded. A test for whiteness is to be found in Chapter 5.

It is possible to apply the KF to a nonlinear (instead of a linear) process model if this model is amenable to linearisation. The resulting filter is commonly called the *Extended Kalman Filter* (Jazwinski, 1970; Anderson and Moore, 1979). Although the Extended Kalman Filter is straightforward to apply, its properties (being a sub-optimal filter) are far from well understood. As a result application of the Extended Kalman filter is a somewhat hit and miss affair in practice.

Many excellent texts (Jazwinski, 1970; Gelb, 1974; Anderson and Moore, 1979; Maybeck, 1982) can be used to supplement this brief introduction to optimal filtering. An interesting collection of articles on the application of the KF are presented in the *IEEE Special Issue on Kalman Filtering* (Sorenson *et al.*, 1983). The KF is a milestone in estimation and has enormous potential in process control applications.

4.4 Steady-state parameter estimation

Having discussed two techniques to estimate unknown constants or variables we return now to one of the problems of process control, namely parameter estimation. In this section parameter estimation using only measurements of the steady state of the process is investigated; the use of transient measurements for parameter estimation is deferred until the next section.

The use of steady-state measurements to obtain the process parameters which determine the process dynamics may, at first, seem impossible. Consider first the noise free case. From Chapter 2 we know that a steady output of the linear state-space model is characterised by

$$y = CA^{-1}Bu \qquad (4.68a)$$

$$= Pu \qquad (4.68b)$$

Stacking each column of P on top of each other into a vector x,

$$x = [P_{11}P_{21} \ldots P_{m1}P_{12} \ldots P_{mr}]^{\mathrm{T}} \qquad (4.69)$$

gives

$$y = C_0 x \qquad (4.70)$$

where C_0 is an $(m \times mr)$ matrix formed from u. To obtain an estimate of x we require (see Section 4.2) more than mr measurements. Let us assume that the process can be operated at M $(> r)$ different steady states (by choosing M different values of u). The parameters in P must not, of course, change. If batch processing is acceptable then the measurement vectors may be stacked on top of each other to give the $(mM \times 1)$ vector

$$z = Hx \qquad (4.71)$$

A least-squares estimator may now be constructed, provided that the different steady-state control signals can be chosen such that H has the rank (mr). The estimator is given by eqn. 4.5:

$$\hat{x} = (H^{T}RH)^{-1}H^{T}Rz \tag{4.5}$$

Now consider the case of noisy steady-state measurements. Instead of eqn. 4.71, the situation is modelled by eqn. 4.30 and under the same condition that H has rank (mr) the Gauss–Markov theorem indicates that the optimum variance estimate of x is given by

$$\hat{x} = (H^{T}V^{-1}H)^{-1}H^{T}V^{-1}z \tag{4.34}$$

Comparing both estimates of x we see that whichever situation we have, a value must be chosen for either R or V. Note how the stochastic viewpoint does not lead to a more complicated estimator than the deterministic (i.e. no noise) case. Even with the 'correct' choice of R or V we have at best estimated only the (mr) parameters of P. There is no way of extricating from P the $n(m + n + r)$ parameters of A, B and C.

Fortunately, two factors in process control problems usually mitigate the unpromising situation sketched above. First, many process model parameters are usually known – they may be physical or chemical constants – and moreover often the same parameter (or a known multiple thereof) appears in a model more than once. For the approach just outlined to work, there must be, of course, no more than mr unknown parameters, and it must be possible to calculate each unknown parameter from P.

Example 4.3

A linear process model with two measured outputs and a single input has the form:

$$A = \begin{bmatrix} -10 & 0 \\ \alpha & -9 \end{bmatrix}; \quad B = \begin{bmatrix} \beta \\ -\beta \end{bmatrix}; \quad C = \begin{bmatrix} 1 & 0 \\ 0 & 1 \end{bmatrix}$$

where α and β are two unknown parameters. Can α and β be estimated from steady-state data? Clearly $mr = 2 =$ number of unknowns. Now since

$$P = -CA^{-1}B = \begin{bmatrix} \beta/10 \\ (\alpha - 10)\beta/90 \end{bmatrix}$$

it is possible, given an estimate of $\beta/10$ and $(\alpha - 10)\beta/90$, to calculate the corresponding α and β from the estimate. If the same process model, but with

$$A = \begin{bmatrix} -9-\alpha & 0 \\ \alpha & -9 \end{bmatrix}$$

had been used it would have proved impossible to estimate α and β.

The second factor which may work in our favour in some problems is that the linear process model is usually derived from a nonlinear mathematical model. The nonlinear relationship between input and output in the steady state may permit parameters to be estimated which would not be possible in the linearised process model.

Example 4.4
A model for the hydrolysis of ester with caustic soda in an ideally mixed reactor is given by the two nonlinear equations

$$\mathring{x}_1(t) = [\beta - x_1(t)]u(t) - \alpha x_1(t) \tag{4.72a}$$

$$\mathring{x}_2(t) = -x_2(t)u(t) + \alpha x_1(t) \tag{4.72b}$$

Both states may be measured. The steady state of the model is given by

$$0 = (\beta - x_1)u - \alpha x_1 \tag{4.73a}$$

$$0 = -x_2 u + \alpha x_1 \tag{4.73b}$$

from which

$$\begin{bmatrix} y_2 u \\ y_1 u \end{bmatrix} = \begin{bmatrix} y_1 & 0 \\ -y_1 & u \end{bmatrix} \begin{bmatrix} \alpha \\ \beta \end{bmatrix} \tag{4.74}$$

so that a least-squares estimate of α and β is possible. Note that if $u = 9$ and $\beta - x_1 = \beta'$, linearisation of eqn. 4.72 gives the system matrices

$$A = \begin{bmatrix} -9 - \alpha & 0 \\ \alpha & -9 \end{bmatrix}, \quad B = \begin{bmatrix} \beta' \\ -\beta' \end{bmatrix}$$

which were used in the previous example.

4.5 Dynamic parameter estimation

The EDS of Chapter 3,

$$x_k = \Phi_k x_k + \Gamma_k u_k + w_k \tag{4.75}$$

$$y_k = C_k x_k + v_k \tag{4.76}$$

contains model parameters Φ_k, Γ_k and C_k and also noise statistics W_k, S_k and V_k that are usually only partially known. If these parameters are time-invariant it may be possible to estimate unknown parameter values using only steady-state measurements as described in the previous section.

There are situations – although experience shows them to be few and far between – where the process under consideration is too little understood for a mathematical model (no matter how crude) to be derived. In that case the EDS equations above are assumed to hold (their generality has already been demonstrated in Chapter 3) and tests are performed on the process so that using dynamic measurements the unknown parameters of the EDS can be estimated. This approach is commonly said to be a *black box* approach.

From the outset it must be said that no technique has yet emerged that can solve the dynamic parameter estimation problem – that of finding Φ_k, Γ_k, C_k, W_k, S_k, and V_k from finite records of u_k and y_k – in a satisfactory way. Clearly the technique should be an optimal one since then questions of existence and stability can be resolved. However, in most cases the state of the process, as well as the parameters of the process, is unknown: this leads to a nonlinear estimation problem (since state variables and parameters are multiplied together in the EDS) for which there is, as yet, no known optimal solution. Therefore, if this, the most straightforward approach, is to be pursued then some compromise must be made either to the optimality or to the generality of the EDS.

The most obvious simplification is to assume that all the states of the EDS can be measured (Mayne, 1963; Eykhoff, 1974). This may seem a very severe restriction, but in some cases it is possible to arrange additional measuring instruments on a temporary basis so that all state variables are available. Let the unknown parameters of Φ_k and Γ_k be arranged into the vectors a_k and b_k, respectively. Then eqn. 4.75 becomes

$$x_{k+1} = X_k a_k + U_k b_k + w_k \tag{4.77a}$$

$$= [X_k U_k]\begin{bmatrix} a_k \\ b_k \end{bmatrix} + w_k \tag{4.77b}$$

where X_k and U_k are matrices of appropriate dimensions containing elements of x_k and u_k, respectively. The number of unknown parameters i.e. $\dim\{a_k\} + \dim\{b_k\}$ must not be greater than the number of equations, which is n. Now if nothing is known concerning the covariance of w_k, or if there is no input noise, the least-squares technique could be applied to eqn. 4.77*b* to estimate a_k and b_k. When W_k is known, eqn. 4.77*b* can be regarded as an output equation for a KF, the state equation of which can be taken to be a random walk process, i.e. with $\{e_k\}$ a zero-mean, independent white noise process,

$$\begin{bmatrix} a_{k+1} \\ b_{k+1} \end{bmatrix} = \begin{bmatrix} a_k \\ b_k \end{bmatrix} + e_k \tag{4.78}$$

Note that whichever method is used, least-squares or minimum variance, the convergence of the estimate (the stability of the estimator) depends critically upon $[X_k U_k]$. Roughly speaking, inputs u_k which guarantee that the estimate converges are called *sufficiently rich* (Moore, 1983). Finally, note that the EDS output equation, eqn. 4.76, can be treated in the same way to obtain estimates of the parameters of the matrix C_k.

An alternative approach is to augment those parts of the state and output equations of the EDS which are known with linear terms containing (unknown) parameter vectors (Schmidt, 1966; Jazwinski, 1970). The assumption that all state variables are measured can then be relaxed, and the KF applied in a straightforward way.

Suboptimal solutions are also possible of course. Since unknown states and parameters are multiplied together one obvious simplification is to linearise the state and output equations of the EDS. Then the Extended Kalman Filter (Section 4.3) can be applied to estimate both states and parameters, suboptimally. Experience has shown that the convergence of the estimates is poor but can be enhanced by modifying the usual EDS form (Ljung, 1981). The poor convergence of the extended Kalman filter in this application has also led to methods which employ the KF in a nearly optimal fashion (Nelson and Stear, 1976; Ahmed, 1983).

Another suboptimal method of parameter estimation is that known as extended least squares. Simultaneous state and parameter estimation is achieved by employing separate state and parameter estimators in a 'bootstrap' fashion; the state estimator uses the best current estimate of the parameters to find $\hat{x}_k$, while the parameter estimator uses the best current state estimate to find $\hat{\Phi}_k$, $\hat{\Gamma}_k$ and $\hat{C}_k$. The convergence properties of the extended least-squares method have been extensively studied (see Anderson and Moore, 1979).

All the foregoing approaches in the parameter estimation problem contain one or more compromises to allow the nonlinear estimation problem to be solved. An important alternative is the following (Mehra, 1970; Martin and Stubberud, 1976; Leondes and Siu, 1980). Recall that the KF operating with correct parameter values results in an innovations sequence that is zero-mean and white (Section 4.3). Starting with an initial guess for the unknown parameters it is possible to adaptively adjust the model parameters until the innovations is zero-mean and white. The model parameters are then the same as those of the process – if it is linear. This technique is claimed to have the ability not only to identify Φ, Γ and C but also W and V.

Stochastic realisation theory (Baram, 1981) also offers an attractive solution to the parameter estimation problem. We have already referred to the realisation theory developed for deterministic systems (Kalman and Ho, 1966). The stochastic realisation theory derives an EDS for a stochastic process by a factorisation of the output covariance. As with the innovations approach, sample covariances replace actual covariances in the computation of the unknown parameters, which are usually assumed to be time-invariant. A drawback is that $V_k = 0$ must be assumed, i.e. there is no measurement noise.

Little has been said, until now, about techniques to obtain the noise statistics, W_k, S_k and V_k of the EDS model. If the model is to be used to design a KF for state estimation then it is actually not necessary to know these noise statistics explicitly. The Kalman gain (but unfortunately not the error covariance) can be computed from (sample) output covariance data. The technique to do this is described in the next section. However, in most practical situations the noise statistics can be treated as variable parameters in the KF design procedure so that their actual values are not critical.

4.6 State estimation

Many problems of process control concern the smoothing, filtering or prediction – in short the estimation – of the state variables of a process. This section considers some of the problems of applying an estimator to a practical situation. Moreover some of the useful tasks that an estimator can perform in process control are listed.

We assume that the process under consideration has been mathematically modelled and that this deterministic model has been transformed into the stochastic EDS. As explained in Chapter 3, the EDS is suitable for estimator (or controller) design, since the estimator will always be implemented by a computer. It will also be assumed that all the model parameters (but not the noise statistics) are either known or have been estimated. Note that as explained in the last chapter, an analogue filter might be necessary to remove high frequency noise.

The choice that must now be made is to whether to apply least-squares or minimum variance estimation. Assuming, as in the majority of applications, an online (or real-time) estimate of the state of the process is required, the estimator of whatever type will have a recursive form. Not so clear is whether a steady-state estimator would suffice in general or whether a complete dynamic estimator is necessary.

A definite answer to these questions is not, at present, possible. The best advice is to 'play safe' where computing speed and memory permit and implement the full KF in preference to either the RLS or steady-state KF, if necessary by artificially assuming the presence of noise. If online performance is satisfactory it is always possible to experiment with the RLS by setting $\Phi_k = I$, $\Gamma_k = W_k = S_k = 0$, $V_k = R_k^{-1}$ and $\bar{x}_0 = 0$ with $\Sigma_0^{-1} = 0$ (see Section 4.3). If the online performance is such that the KF gain very quickly reaches a steady value maybe the simplification of using the steady-state KF justifies its implementation.

One of the major practical difficulties in KF design is that the noise statistics (i.e. W_k, S_k and V_k) are usually unknown. If plant data is available then this difficulty can be overcome in a quite satisfactory way to be described shortly. Otherwise values of W_k would be guessed (S_k is usually set to zero) and simulations used to explore the KF response and sensitivity to chosen values. If V_k is known it may be a singular matrix, implying that certain observations are not corrupted by measurement noise. We show in this section how to disentangle uncorrupted from noisy states.

The structure of the KF as given in Section 4.3 is not the most suited to numerical computations. The so-called square-root filter is therefore introduced. Then various derivatives of the KF, suitable for smoothing and prediction, are discussed. Finally some useful applications for estimation in process control are listed.

4.6.1 *Estimator design from covariance data*

We begin by showing how it is possible to design a KF without explicit knowledge of the noise statistics (Mehra, 1970; Anderson and Moore, 1979). Since the noise terms often account for modelling errors, truncation errors etc. of an ill-defined nature (recall the discussion in Chapter 3) this capability is extremely important. The technique to be described is based upon the use of the output covariance $E\{y_k y_l^{\mathrm{T}}\}$; it will be assumed henceforth that measurements have been performed (*a priori*) on the process to provide this quantity.

The process model is the familiar EDS:

$$x_{k+1} = \Phi_k x_k + \Gamma_k u_k + w_k \tag{4.79}$$

$$y_k = C_k x_k + v_k \tag{4.80}$$

where the usual assumptions concerning correlation hold and

$$E\{w_k w_l^{\mathrm{T}}\} = W_k \delta_{kl} \tag{4.81a}$$

$$E\{w_k v_l^{\mathrm{T}}\} = S_k \delta_{kl} \tag{4.81b}$$

$$E\{v_k v_l^{\mathrm{T}}\} = V_k \delta_{kl} \tag{4.81c}$$

The model parameters of eqns. 4.79 and 4.80 are known (perhaps from physical or chemical balances) while the noise statistics, eqn. 4.81, are unknown. First, an expression for the output covariance $E\{y_k y_l^{\mathrm{T}}\}$ is derived for the situation where $u_k \equiv 0$ (the simplest case). Then it is shown how the Kalman gain, K_k, can be calculated from the output covariance.

For $k > l$ and with $u_k = 0$, eqn. 4.79 may be written

$$x_k = \psi_{k,l} x_l + \sum_{i=l}^{k-1} \psi_{k,i+1} w_i \tag{4.82}$$

where we have defined

$$\psi_{k,l} \triangleq \Phi_k \Phi_{k-1} \ldots \Phi_l \tag{4.83}$$

It follows that

$$\begin{aligned} E\{x_k v_l^{\mathrm{T}}\} &= E\left\{\psi_{k,l} x_l v_l^{\mathrm{T}} + \sum_{i=1}^{k-1} \psi_{k,i+1} w_i v_l^{\mathrm{T}}\right\} \\ &= E\left\{\sum_{i=1}^{k-1} \psi_{k,i+1} w_i v_l^{\mathrm{T}}\right\} \\ &= \psi_{k,l+1} S_l \end{aligned} \tag{4.84}$$

From eqn. 4.82 the state covariance is easily derived:

$$\begin{aligned} E\{x_k x_l^{\mathrm{T}}\} &= E\left\{\psi_{k,l} x_l x_l^{\mathrm{T}} + \sum_{i=l}^{k-1} \psi_{k,i+1} w_i x_l^{\mathrm{T}}\right\} \\ &= \psi_{k,l} E\{x_l x_l^{\mathrm{T}}\} \end{aligned} \tag{4.85}$$

Now the output covariance is found from eqn. 4.80 to be

$$E\{y_k y_l^T\} = C_k E\{x_k x_l^T\} C_l^T + C_k E\{x_k v_l^T\} + E\{v_k x_l^T\} C_l^T + E\{v_k v_l^T\} \tag{4.86}$$

and since $k > l$ the last two terms are zero. On substituting eqns. 4.84 and 4.85 and defining

$$P_l \triangleq E\{x_l x_l^T\} \tag{4.87}$$

as the state variance, the output covariance may be written as

$$\begin{aligned} E\{y_k y_l^T\} &= C_k \psi_{k,l} P_l C_l^T + C_k \psi_{k,l+1} S_l \\ &= C_k \psi_{k,l+1}(\Phi_l P_l C_l^T + S_l) \end{aligned} \tag{4.88}$$

The bracketed quantity we define as M_l, i.e.

$$M_l \triangleq \Phi_l P_l C_l^T + S_l \tag{4.89}$$

Since the Kalman gain can be expressed in terms of this quantity, we require an explicit relationship relating this to measurable quantities. Recall that if (C_k, Φ_k) is observable (not necessarily uniformly) then for all k and some $j > 0$

$$\sum_{i=k}^{k+j-1} \psi_{i,k}^T C_i^T C_i \psi_{i,k} > 0 \tag{4.90}$$

This suggests forming from eqn. 4.88 the expression

$$\psi_{k,l+1}^T C_k^T E\{y_k y_l^T\} = \psi_{k,l+1}^T C_k^T C_k \psi_{k,l+1} M_l \tag{4.91}$$

Choose now $k = k_1, k_2, \ldots, k_p$, all $> l$. Adding each of the expressions together gives

$$\sum_{i=k_1}^{k_p} \psi_{i,l+1}^T C_i^T E\{y_i y_l^T\} = \left(\sum_{i=k_1}^{k_p} \psi_{i,l+1}^T C_i^T C_i \psi_{i,l+1}\right) M_l \tag{4.92}$$

so that if the process is observable there must exist a $k_p > l$ such that the term in brackets is invertible. Thus, for some k_p,

$$M_l = \left(\sum_{i=k_1}^{k_p} \psi_{i,l+1}^T C_i^T C_i \psi_{i,l+1}\right)^{-1} \sum_{i=k_1}^{k_p} \psi_{i,l+1}^T C_i^T E\{y_i y_l^T\} \tag{4.93}$$

which is an explicit expression for M_l in terms of given or measurable quantities.

Let the output variance, $E\{y_k y_k^T\}$, also be measured and denote this by

$$L_k \triangleq E\{y_k y_k^T\} \tag{4.94}$$

Then it can be shown (Son and Anderson, 1973) that the Kalman gain is given by

$$K_k = -(\Phi_k T_k C_k^T - M_k)(L_k - C_k T_k C_k^T)^{-1} \tag{4.95}$$

where

$$T_{k+1} = \Phi_k T_k \Phi_k^{\mathrm{T}} + (\Phi_k T_k C_k^{\mathrm{T}} - M_k)(L_k - C_k T_k C_k^{\mathrm{T}})^{-1} \times (\Phi_k T_k C_k^{\mathrm{T}} - M_k)^{\mathrm{T}} \tag{4.96}$$

initialised by $T_0 = 0$.

Together, eqns. 4.95 and 4.96 permit the Kalman gain, K_k, to be calculated if the output covariance is known for all k and l. This requirement precludes the straightforward application of this technique to online estimation. However, there are many practical situations where the sampling rate can be made much higher than the desired frequency of estimation; in that case eqn. 4.92 could, after a delay, provide values of M_k to mechanise the Kalman filter gain equation, eqn. 4.95. The error covariance of the KF can never be precomputed since it depends on P_k (Anderson and Moore, 1979). If, for some reason, $u_k \neq 0$, then it can easily be shown that the analysis still holds but the results are more complex: in that case the mean of the output, $E\{y_k\}$, must be measured as well as the output covariance. Finally it should be noted that the process has been assumed to be completely observable. If this is not the case then it may well indicate that more, or different, sensors should be applied to the process.

4.6.2 *Noise-free estimation*

There are strong reasons for assuming the presence of white process noise and white measurement noise in the EDS which is to be used as the process model for estimator design. Some of these reaasons have already been expounded in Chapter 3. Moreover, it can be shown (Anderson and Moore, 1979) that mild nonlinearities can be modelled for the purposes of KF design by incorporating additional process and measurement noise. That the noise is taken to be white represents a sort of 'worst case' design condition, and besides, the improvement to be found by modelling coloured noise instead is often marginal (Kwakernaak and Sivan, 1972). Note that the KF is applicable to processes with coloured noise (usually at the expense of an increase in the dimensionality of the filter) or bounded section nonlinearities (Anderson and Moore, 1979).

A situation of more practical importance and frequently encountered in process control is when some of the observations are completely free from measurement noise. In view of what has just been said it is reasonable to artificially induce measurement noise in the EDS by choosing $V_k > 0$ (how this can be achieved depends upon the particular situation) and proceeding with the normal KF design. For certain applications such an approach might lead to an undesirable filter response, although it is difficult to be more specific than to suggest simulation of the estimator and process as the only true test. Suppose we were to proceed with the KF design when $V_k \geqslant 0$, i.e. when some observations are free from measurement noise. Looking at both the error covariance and filter gain equations, eqns. 4.53 and 4.52, the quantity

$$C_k \Sigma_{k/k-1} C_k^{\mathrm{T}} + V_k$$

must be inverted. The first term is generally only non-negative definite so that the inversion will fail unless V_k is positive definite for all k, see Section 4.3. Therefore, when some of the measurements are noise-free the following technique should be applied to produce state estimates.

Recalling that there are m measurements, let m_0 of these be noise-free, with $m_0 < m$. For the process model we take the EDS, eqns. 4.79 to 4.81 but with the output equation time-invariant, i.e. $C_k = C$ and $V_k = V$ (stationary measurement noise). This assumption simplifies our presentation but is not essential (Tse and Athans, 1970). If T is any $(m \times m)$ invertible matrix, then eqn. 4.80 may be written

$$\tilde{y}_k = TCx_k + \tilde{v}_k \tag{4.97}$$

where $\tilde{y}_k = Ty_k$ and $\tilde{v}_k = Tv_k$. Now

$$E\{\tilde{v}_k\tilde{v}_k^{\mathrm{T}}\} = TVT^{\mathrm{T}} \tag{4.98}$$

and since V is a non-negative definite symmetric matrix there exists an invertible matrix T such that (see Ayres, 1974)

$$TVT^{\mathrm{T}} = \begin{bmatrix} 0 & 0 \\ 0 & I_{m-m_0} \end{bmatrix} \tag{4.99}$$

Hence the first m_0 entries of $\tilde{v}_k$ are zero-mean white noise processes with zero variance, i.e. are zero. We must now try to isolate the corresponding state variables, since these can be measured without noise. The similarity transformation (Chapter 1) $\tilde{x}_k = Hx_k$, where H is an $(n \times n)$ invertible matrix, is applied to eqn. 4.97 to give

$$\tilde{y}_k = TCH^{-1}\tilde{x}_k + \tilde{v}_k \tag{4.100}$$

Now

$$TC = [I_m \quad 0]\begin{bmatrix} TC \\ \tilde{H} \end{bmatrix} \tag{4.101}$$

where the $(n - m \times n)$ matrix $\tilde{H}$ may be chosen (T has full rank and C has rank m if none of the measurements is redundant) such that

$$\begin{bmatrix} TC \\ \tilde{H} \end{bmatrix} \triangleq H \tag{4.102}$$

is an invertible matrix. With this definition it is obvious from eqn. 4.101 that

$$TCH^{-1} = [I_m \quad 0] = \begin{bmatrix} I_{m_0} & 0 & 0 \\ 0 & I_{m-m_0} & 0 \end{bmatrix} \tag{4.103}$$

so that from the transformed output eqn. 4.100 the first m_0 measurements of $\tilde{y}_k$ equal the first m_0 (noise-free) state variables of $\tilde{x}_k$. The remaining $n - m_0$ state variables can be estimated by using the KF on the process model:

$$\tilde{x}_{k+1} = H\Phi_k H^{-1}\tilde{x}_k + H_k\Gamma u_k + \tilde{w}_k \tag{4.104}$$

$$\tilde{y}_k = [0_{m_0} I_{m-m_0} 0]\tilde{x}_k + \bar{v}_k \tag{4.105}$$

$$E[\tilde{w}_k \tilde{w}_l^{\mathrm{T}}] = HW_k H^{\mathrm{T}}\delta_{kl} \tag{4.106a}$$

$$E[\tilde{w}_k \bar{v}_l^{\mathrm{T}}] = HS_k T^{\mathrm{T}} \begin{bmatrix} 0_{m_0} \\ I_{m-m_0} \end{bmatrix} \delta_{kl} \tag{4.106b}$$

$$E[\bar{v}_k \bar{v}_l^{\mathrm{T}}] = I_{m-m_0}\delta_{kl} \tag{4.106c}$$

4.6.3 *Square-root filter*

It is well known and documented that implementation of the KF equations is not, in practice, a straightforward matter. The equations are sensitive to roundoff errors, and poor numerical accuracy often results. For example, computation of the error covariance $\Sigma_{k/k-1}$ from the error covariance equation, eqn. 4.53, may result in a matrix which is not non-negative definite. Potter's *square-root filter* (Bierman, 1977) has been successfully demonstrated to combat these numerical difficulties. In essence, this filter is formulated in terms of the square root of the error covariance matrix; multiplication of the square root (which is not necessarily non-negative definite) with its transpose must then lead to a non-negative definite error covariance matrix. Recall that the square root of a matrix has already been discussed (Chapter 1). A brief description of the square-root filter (Morf and Kailath, 1975) will now be given. It should be born in mind that apart from its demonstrated numerical supremacy, the square-root filter and the Kalman filter derive from the same equations and (theoretically) produce the same state estimate.

Let A and B be any matrices, and let T be an orthogonal matrix ($T^{\mathrm{T}}T = I$) of appropriate dimensions such that

$$TA = B \tag{4.107}$$

Then

$$A^{\mathrm{T}}T^{\mathrm{T}}TA = B^{\mathrm{T}}B$$

$$A^{\mathrm{T}}A = B^{\mathrm{T}}B \tag{4.108}$$

Now if the KF error covariance equation, eqn. 4.53, could be arranged in the form of this last equation then its square root, eqn. 4.107, could be computed instead. From eqn. 4.53

$$\Phi_k\Sigma_{k/k-1}\Phi_k^{\mathrm{T}} + W_k = \Sigma_{k+1/k} + (\Phi_k\Sigma_{k/k-1}C_k^{\mathrm{T}} + S_k) \times (C_k\Sigma_{k/k-1}C_k^{\mathrm{T}} + V_k)^{-1} \times (\Phi_k\Sigma_{k/k-1}C_k^{\mathrm{T}} + S_k)^{\mathrm{T}} \tag{4.109}$$

and this suggests the following form for the $(2n + m \times mn)$ matrix A:

$$A = \begin{bmatrix} V_k^{\frac{1}{2}} & V_k^{-\frac{1}{2}} S_k^{\mathrm{T}} \\ \Sigma_{k/k-1}^{\frac{1}{2}} C_k^{\mathrm{T}} & \Sigma_{k/k-1}^{\frac{1}{2}} \Phi_k^{\mathrm{T}} \\ 0 & W_k^{\frac{1}{2}} \end{bmatrix} \tag{4.110}$$

where it has been assumed that $V_k > 0$, and that square roots can be found numerically, by using special techniques such as the Cholesky decomposition (Bierman, 1977). Note how all the terms in A are known at time kT_s, and that A is the square root of the matrix

$$A^{\mathrm{T}}A = \begin{bmatrix} V_k + C_k \Sigma_{k/k-1} C_k^{\mathrm{T}} & S_k^{\mathrm{T}} + C_k \Sigma_{k/k-1} \Phi_k^{\mathrm{T}} \\ S_k + \Phi_k \Sigma_{k/k-1} C_k^{\mathrm{T}} & S_k V_k^{-1} S_k^{\mathrm{T}} + \Phi_k \Sigma_{k/k-1} \Phi_k^{\mathrm{T}} + W_k \end{bmatrix} \tag{4.111}$$

If the matrix $\tilde{K}_k$ is now introduced, where

$$\tilde{K}_k \triangleq (\Phi_k \Sigma_{k/k-1} C_k^{\mathrm{T}} + S_k)(C_k \Sigma_{k/k-1} C_k^{\mathrm{T}} + V_k)^{-\frac{1}{2}} \tag{4.112}$$

then a possible choice for the $(2m + n \times mn)$ matrix B is

$$B = \begin{bmatrix} (V_k + C_k \Sigma_{k/k-1} C_k^{\mathrm{T}})^{\frac{1}{2}} & \tilde{K}_k^{\mathrm{T}} \\ 0 & \Sigma_{k+1/k}^{\frac{1}{2}} \\ 0 & V_k^{-\frac{1}{2}} S_k^{\mathrm{T}} \end{bmatrix} \tag{4.113}$$

since

$$B^{\mathrm{T}}B = \begin{bmatrix} V_k + C_k \Sigma_{k/k-1} C_k^{\mathrm{T}} & (V_k + C_k \Sigma_{k/k-1} C_k^{\mathrm{T}})^{\frac{1}{2}} \tilde{K}_k^{\mathrm{T}} \\ \tilde{K}_k (V_k + C_k \Sigma_{k/k-1} C_k^{\mathrm{T}})^{\frac{1}{2}} & S_k V_k^{-1} S_k^{\mathrm{T}} + \tilde{K}_k \tilde{K}_k^{\mathrm{T}} + \Sigma_{k+1/k} \end{bmatrix} \tag{4.114}$$

which in view of eqn. 4.109 is $A^{\mathrm{T}}A$. If B could be found from A then not only is the square root of the error covariance, $\Sigma_{k+1/k}$, immediately available, but the estimate of the state, eqns. 4.50 and 4.52, could be computed from some of the remaining terms of B, since

$$\hat{x}_{k+1/k} = \Phi_k \hat{x}_{k/k-1} + \Gamma_k u_k + \tilde{K}_k (C_k \Sigma_{k/k-1} C_k^{\mathrm{T}} + V_k)^{-\frac{1}{2}} (y_k - C_k \hat{x}_{k/k-1}) \tag{4.115}$$

One method of computing B from A, using eqn. 4.107, is to use the Householder transformation triangularisation (Bierman, 1977) which results in a B having the same structure as in eqn. 4.113 and with the term

$$(V_k + C_k \Sigma_{k/k-1} C_k^{\mathrm{T}})^{\frac{1}{2}}$$

in upper triangular form. The latter is useful since the inverse can easily then be computed for insertion into eqn. 4.115. A check that the B matrix has been found correctly is that its bottom right hand term, $V_k^{-\frac{1}{2}} S_k^{\mathrm{T}}$, is the same as that appearing as the top right right hand term in A.

Of course, if V_k is only non-negative definite and not positive definite, the inverse of V_k cannot be found and the above analysis breaks down. However, it should be recalled that for a well-conditioned (i.e. bounded error and stable) KF, $V_k \geqslant \alpha I > 0$. The original analysis (Morf and Kailath, 1975) assumed that $W_k > 0$, which is clearly a stronger assumption than is made here. Note finally that owing to the similarity of the RLS and KF, the square-root filter is also applicable whenever the RLS must be implemented.

4.6.4 *Prediction*

Among the possible process control applications of the KF – a discussion of which concludes this chapter – are smoothing and prediction. Now we have already remarked that the KF presented so far is actually not a filter but a (one-step) predictor in that it predicts the state one time-step ahead of the available measurements. The true filter equations (see Anderson and Moore, 1979) turn out to be more complicated than the KF equations given here; finding the true filtered estimates rarely justifies the extra effort required, so no more will be said about that topic. What is more interesting are the estimates $\hat{x}_{k+N/k}$ and $\hat{x}_{k-N/k}$ which will now be derived.

We begin with prediction, or more correctly with finding the N-step predictor, and showing how the prediction $\hat{x}_{k+N/k}$, where $N = 1, 2, \ldots$, is related to the one-step prediction $\hat{x}_{k+1/k}$. By successive application of the state equation, eqn. 4.79, it is easily shown that

$$x_{k+N} = [\Phi_{k+N-1} \ldots \Phi_{k+2}\Phi_{k+1}]x_{k+1} + \sum_{i=k+1}^{k+N-1} [\Phi_{k+N-1} \ldots \Phi_{i+2}\Phi_{i+1}][\Gamma_i u_i + w_i] \tag{4.116}$$

Taking the conditional expectation of both sides of this equation yields the N-step predictor:

$$\hat{x}_{k+N/k} = [\Phi_{k+N-1} \ldots \Phi_{k+2}\Phi_{k+1}]\hat{x}_{k+1/k} + \sum_{i=k+1}^{k+N-1} [\Phi_{k+N-1} \ldots \Phi_{i+2}\Phi_{i+1})\Gamma_i u_i \tag{4.117}$$

The N-step predictor is equivalently given by the recursive relationship

$$\hat{x}_{k+i+1/k} = \Phi_{k+i}\hat{x}_{k+i/k} + \Gamma_{k+i}u_{k+i} \tag{4.118}$$

initialised at $i = 1$ by the one-step prediction, $\hat{x}_{k+1/k}$ available from the KF. It has the properties associated with any linear dynamic system and it is easy to show (see Anderson and Moore, 1979) that, at least when $W_k \neq 0$, the error covariance grows larger and larger the further one gets away from the time at which the measurements were made – in accordance with our intuition.

4.6.5 *Smoothing*

Smoothing is concerned with obtaining estimates of previous states; in particular (fixed-point smoothing) we wish to obtain $\hat{x}_{k-N/k}$, where $N = 1, 2, \ldots$.

To this end, define a new state vector, z_k, by

$$z_{k+1} = x_{k-N} = z_k \tag{4.119}$$

for some given N and all k. It follows that

$$\hat{z}_{k+1/k} = x_{k-N/k} \tag{4.120}$$

is the optimal smoothed estimate of x_{k-N}. The optimal smoother may be developed (Anderson and Moore, 1979) by augmenting the state of the usual process model with z_k:

$$\begin{bmatrix} x_{k+1} \\ z_{k+1} \end{bmatrix} = \begin{bmatrix} \Phi_k & 0 \\ 0 & I \end{bmatrix} \begin{bmatrix} x_k \\ z_k \end{bmatrix} + \begin{bmatrix} \Gamma_k \\ 0 \end{bmatrix} u_k + w_k \tag{4.121a}$$

$$y_k = [C_k \quad 0] \begin{bmatrix} x_k \\ z_k \end{bmatrix} + v_k \tag{4.121b}$$

$$E\{w_k w_l^{\mathrm{T}}\} = \begin{bmatrix} W_k & 0 \\ 0 & 0 \end{bmatrix} \delta_{kl} \tag{4.122a}$$

$$E\{w_k v_l^{\mathrm{T}}\} = \begin{bmatrix} S_k \\ 0 \end{bmatrix} \delta_{kl} \tag{4.122b}$$

$$E\{v_k v_l^{\mathrm{T}}\} = V_k \delta_{kl} \tag{4.122c}$$

Then, from eqn. 4.50, the KF is given by

$$\begin{bmatrix} \hat{x}_{k+1/k} \\ \hat{z}_{k+1/k} \end{bmatrix} = \begin{bmatrix} \Phi_k - K_k C_k & 0 \\ -L_k C_k & I \end{bmatrix} \begin{bmatrix} \hat{x}_{k/k-1} \\ \hat{z}_{k/k-1} \end{bmatrix} + \begin{bmatrix} K_k \\ L_k \end{bmatrix} y_k + \begin{bmatrix} \Gamma_k \\ 0 \end{bmatrix} u_k \tag{4.123}$$

so

$$\hat{z}_{k+1/k} = \hat{z}_{k/k-1} + L_k(y_k - C_k \hat{x}_{k/k-1}) \tag{4.124}$$

where the Kalman gain matrix, $[K_k^{\mathrm{T}} L_k^{\mathrm{T}}]^{\mathrm{T}}$, is given (eqn. 4.52) by

$$\begin{bmatrix} K_k \\ L_k \end{bmatrix} = \left(\begin{bmatrix} \Phi_k & 0 \\ 0 & I \end{bmatrix} \begin{bmatrix} \Sigma_{k/k-1} \Sigma_{k/k-1}^{z\mathrm{T}} \\ \Sigma_{k/k-1}^{z} \Sigma_{k/k-1}^{zz} \end{bmatrix} \begin{bmatrix} C_k^{\mathrm{T}} \\ 0 \end{bmatrix} + \begin{bmatrix} S_k \\ 0 \end{bmatrix} \right) (C_k \Sigma_{k/k-1} C_k^{\mathrm{T}} + V_k)^{-1} \tag{4.125}$$

A useful form of error covariance equation, eqn. 4.53, is found from substituting eqn. 4.52 into eqn. 4.53:

$$\begin{aligned} \Sigma_{k+1/k} &= \Phi_k \Sigma_{k/k-1} \Phi_k^{\mathrm{T}} - K_k(\Phi_k \Sigma_{k/k-1} C_k^{\mathrm{T}} + S_k)^{\mathrm{T}} + W_k \\ &= (\Phi_k - K_k C_k) \Sigma_{k/k-1} \Phi_k^{\mathrm{T}} - K_k S_k^{\mathrm{T}} + W_k \end{aligned} \tag{4.126}$$

Expressing the error covariance of the augmented system by means of the above form, it can be shown that

$$\Sigma^z_{k+1/k} = \Sigma^z_{k/k-1}(\Phi^T_k - C^T_k K^T_k) - L_k S^T_k \tag{4.127}$$

with

$$K_k = (\Phi_k \Sigma_{k/k-1} C^T_k + S_k)(C_k \Sigma_{k/k-1} C^T_k + V_k)^{-1} \tag{4.128}$$

and

$$L_k = \Sigma^z_{k/k-1} C^T_k (C_k \Sigma_{k/k-1} C^T_k + V_k)^{-1} \tag{4.129}$$

We see that the optimal smoother, given by eqn. 4.124, has the same dimension as the KF and is driven by the KF innovations. The smoother gain matrix, L_k, is computed from eqns. 4.126 to 4.129. We can therefore appreciate that smoothing is an extension of filtering: given the KF of a process we can attach the optimal smoother, eqn. 4.124, driven by the KF innovations, to obtain $\hat{x}_{k-N/k}$.

4.6.6 *Applications of estimation in process control*

Lastly, we consider applications. It must first be stated that compared to the potential number of applications, only relatively few examples of estimators are to be found in the process industries. There has been a much greater concentration of effort upon applying advanced control strategies than on estimation, and quite unjustifiably so. This is because a good controller will invariably require state estimates; without them it will usually malfunction.

In the process industry the scope for estimation and application of the KF is considerable. Here are some examples.

1 State estimation for process control (see Chapter 6).
2 State estimation of variables which are inaccessible to measuring instruments.
3 State estimation of variables in environments likely to corrode or damage measuring instruments.
4 Smoothing of measurement data to deduce the initial concentrations of reactants present in batch reactors.
5 Prediction of unsafe process conditions.
6 Sensor failure detection.
7 Estimation of process apparatus parameters such as pump coefficients, friction factors and valve characteristics.

References

AHMED, M. S. (1983): 'On bootstrap estimation of system parameters and states in linear systems', *IEEE Trans. Auto. Control*, **AC-28**, pp. 805–806

AYRES, F. (1974): 'Matrices' (McGraw-Hill)

ANDERSON, B. D. O. and MOORE, J. B. (1979): 'Optimal filtering' (Prentice-Hall)

ANDERSON, B. D. O. and MOORE, J. B. (1981): 'Detectability and stabilizability of time-varying discrete-time linear systems', *SIAM J. Control & Opt.*, **19**, 1, pp. 20–32

BARAM, Y. (1981): 'Realization and reduction of markovian models from nonstationary data', *IEEE Trans. Auto. Control*, **AC-26**, pp. 1225–1231

BARD, Y. (1974): 'Nonlinear parameter estimation' (Academic Press)

BIERMAN, G. J. (1977): 'Factorization methods for discrete sequential estimation' (Academic Press)

CLEMENTS, D. J. and ANDERSON, B. D. O. (1978): 'Singular optimal control: the linear-quadratic problem' (Springer-Verlag)

EYKHOFF, P. (1974): 'System identification' (John Wiley)

GELB, A., ed. (1974): 'Applied optimal estimation' (MIT Press)

GOODWIN, G. C. and PAYNE, R. L. (1977): 'Dynamic system identification: experiment design and data analysis' (Academic Press)

HASTINGS-JAMES, R. and SAGE, M. W. (1969): 'Recursive generalised-least-squares procedure for on-line identification of process parameters', *Proc. IEE*, **116**, pp. 2057–2062

JAZWINSKI, A. H. (1970): 'Stochastic processes and filtering theory' (Academic Press)

JUNKINS, J. L. (1978): 'An introduction to optimal estimation of dynamical systems' (Sijthoff and Noordhoff, The Netherlands)

KALMAN, R. E. and HO, B. L. (1966): 'Effective construction of linear state-variable models from input/output functions', *Regelungstechnik*, **14**, pp. 545–548

KALMAN, R. E. (1960): 'A new approach to linear filtering and prediction problems', *J. Basic Eng.*, Trans. ASME, Series D, **82**, 1, pp. 35–45

KALMAN, R. E. (1978): 'A retrospective after twenty years: from the pure to the applied', Proc. AGU Chapman Conference, Pittsburgh, pp. 31–53

KWAKERNAAK, H. and SIVAN, R. (1972): 'Linear optimal control systems' (John Wiley)

LAWSON, C. L. and HANSON, R. J. (1974): 'Solving least-squares problems' (Prentice Hall)

LEONDES, C. T. and SIU, T. K. (1980): 'Identification both of the unknown plant and noise parameters of the Kalman Filter', *Int. J. Sys. Sci.*, **11**, 6, pp. 711–720

LJUNG, L. (1981): 'Recursive identification. In stochastic systems: the mathematics of filtering and identification and applications' (D. Reidel Co.)

LUENBERGER, D. G. (1964): 'Observing the state of a linear system', *IEEE Trans. Mil. Electron.*, **8**, pp. 74–80

MAYBECK, P. S. (1982): 'Stochastic models, estimation and controls. Vols. I–III' (Academic Press)

MAYNE, D. Q. (1963): 'Optimal non-stationary estimation of the parameters of a linear system with gaussian inputs', *J. Elec. Cont.*, **14**, pp. 101–112

MARTIN, W. C. and STUBBERUD, A. R. (1976): 'An uncoupling method for linear system identification', *IEEE Trans. Auto. Control*, **AC-21**, pp. 506–509

MEHRA, R. K. (1970): 'On the identification of variances and adaptive Kalman filtering', *IEEE Trans. Auto. Control*, **AC-15**, pp. 175–184

MOORE, J. B. (1983): 'Persistence of excitation in extended least squares', *IEEE Trans. Auto. Control*, **AC-28**, 1, pp. 60–68

MORF, M. and KAILATH, T. (1976): 'Square-root algorithms for least-squares estimation', *IEEE Trans. Auto. Control*, **AC-20**, 4, pp. 487–497

NELSON, L. W. and STEAR, E. (1976): 'The simultaneous on-line estimation of parameters and states in linear systems', *IEEE Trans. Auto. Control*, **AC-21**, pp. 94–98

SCHMIDT, S. F. (1966): 'Application of state-space methods to navigation problems', *Advan. Control Systems*, **3**, pp. 293–340

SMITH, G. L. (1965): 'The scientific interential relationships between statistical estimation, decision theory and modern filter theory', *Proc. JACC*, pp. 350–359

SON, L. H. and ANDERSON, B. D. O. (1973): 'Design of Kalman filters using signal model output statistics', *Proc. IEE*, **120**, 2, pp. 312–318

SORENSON, H. W. (1980): 'Parameter estimation: principles and problems' (M. Dekker Inc.)
SORENSON, H. W. *et al.* (1983): 'Special issue on application of Kalman filtering', *IEEE Trans. Auto. Control*, **AC-28**, 3, pp. 253–436
TSE, E. and ATHANS, M. (1970): 'Optimal minimal-order observer estimators for discrete linear time-varying systems', *IEEE Trans. Auto. Control*, **AC-15**, 4, pp. 416–426

Chapter 5

Simulation

5.1 Introduction

A simulation of a process is the display of the solutions of a model of that process to various test inputs (forcing functions) or initial conditions so chosen that various 'worse-case' scenarios can be examined. Often step functions (Chapter 1) are chosen to perturb the model, as in Sections 2.2.5 and 3.3: these step changes in input represent a far more stringent test for the model than is ever likely to occur to the process in practice. The type of simulation of particular interest is digital simulation, where a digital computer and associated peripherals are used for the solution and display task. Analogue or hybrid simulations should only be considered in exceptional circumstances and then, in view of their time-consuming nature, are best left to the specialist.

The simulation model may be either a continuous-time model (Chapter 2), or, if the model represents a sample-data system, a discrete-time model (Chapter 3). Furthermore, the model may be nonlinear or linear, deterministic or sometimes stochastic, and be finite or infinite dimensional (Section 2.2.1). In this latter case we assume that the spatial co-ordinate(s) of the distributed parameter model has been discretised.

Simulation packages, such as CSMP, MIMIC and GSSP, are available and can sometimes save time in terms of the total time taken for a particular study. This is because the user need not be concerned with the display aspect of simulation (tables, charts and graphics output being part of the simulation package) and may choose from many equation solving techniques depending on the problem at hand. Moreover simulation packages are useful during the mathematical modelling stage of a study since they permit the user to select and try any number of simple models, such as described in Section 2.2.5. A disadvantage, however, is that the general nature of the simulation package accrues a large overhead in both software and descriptive literature.

At this point it is necessary to state what sort of simulation is of concern to the process control engineer. We are not involved, for example, in simulations of large or very large (i.e. hundreds of variables) scale systems such as oil reservoirs or complete chemical plants. Not that there are no control or estimation problems in these fields, but they occur less frequently. Special techniques (see, for example, Davison, 1973; Enright, 1979) have been developed for this type of simulation, but are not described here. Some simulations that concern us involve *stiff systems*. For process control simulations, however, we can state that simulation run times encountered are seldom a point of concern. It may also be remarked that the techniques of simulation and the simulation packages available are now so well developed that when used with modern computers the problems of attaining sufficient accuracy in the simulation become insignificant when compared to the inaccuracies of mathematical modelling. The related topics of roundoff, truncation and accumulation errors are well documentred in books on numerical analysis (see, for example, Scheid, 1968).

The remainder of this chapter is organised as follows. A brief motivation which captures the chief applications of simulation for our purposes will be presented next. Then comes a section on steady state simulation, or how to solve implicit algebraic equations. This is followed by dynamic simulation with the well known Runge–Kutta procedure. The final section reviews the simulation of stiff differential equations.

5.2 Applications of simulation in process control

Simulation in process control should complement the theoretical analysis presented of system design and behaviour. Why it is so important can be seen from the following list which catalogues the principal applications of simulation in process control:

1 To compare the nonlinear model with the actual process it describes. Simulating the process (possibly slower than real time) can increase process operating skills and knowledge of the process model, including those phenomena not reproduced by linear models such as limit cycling, saturation, hysteresis etc.
2 To examine the useful range of approximation introduced by the linearised model, by comparing its behaviour with that of the nonlinear model describing the process.
3 To investigate the robustness of the linear model, i.e. the sensitivity of the properties of the model to parametric changes.
4 To check the working of both controllers and estimators, designed using linear process models, on the nonlinear model prior to installation.
5 To check the implementation of controller and estimator algorithms in computers and microprocessors. Integer arithmetic and the number of bits per byte of ADC (see Chapter 3) can greatly influence the controller or estimator performance.

Example 5.1
Fig. 5.1*a* is a simulation of the (ideal) response of an optimally controlled industrial process to a set-point change in temperature. Fig. 5.1*b* is the same response, but with the 8-bits analogue-to-digital temperature conversion and the integer arithmetic of the microprocessor which implements the optimal controller also simulated. Notice that although the temperature response (T) is unchanged, the control signal (u) is unfavourably influenced. Possible remedies for this sort of behaviour of u_k include more accurate ADC, floating-point (instead of integer) processing and decreasing the range of the temperature sensor by hardware switching.

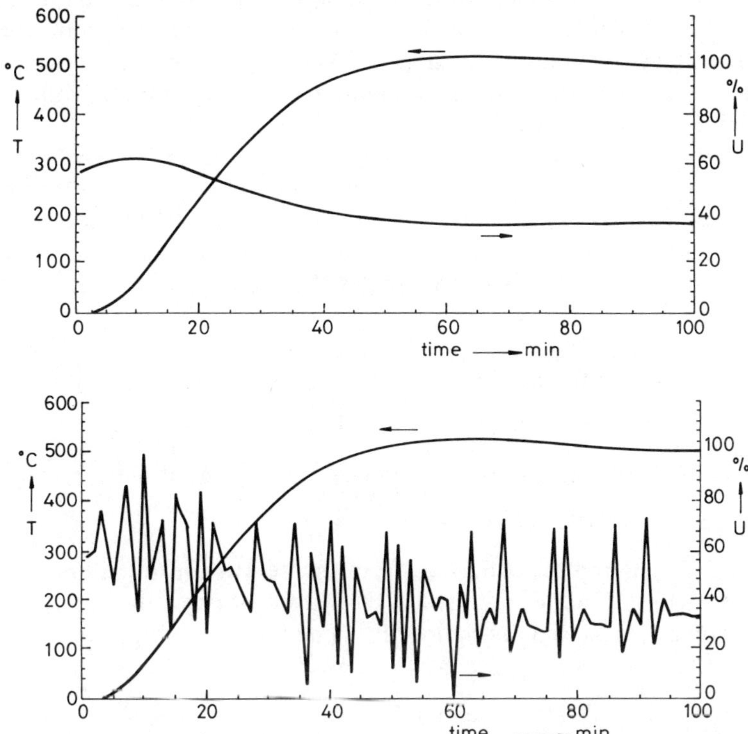

Fig. 5.1 *Responses to setpoint change in temperature*
a With optimal controller
b With optimal controller, 8 bits ADC and integer arithmetic

5.3 Steady-state simulation

Mathematical models are sometimes constructed using mathematical functions such as the exponential, polynomial etc. to describe observed data such as growth curves, temperature profiles etc. Because of the presence of these functions it may not be possible to obtain an explicit relationship describing the steady state of the model. In this section some techniques are presented which may be applied to solve implicit nonlinear algebraic equations of the form:

$$0 = f(x) \tag{5.1}$$

where f is a vector of deterministic functions and x is the steady state of the model.

Steady-state simulation (or process flowsheeting) has been extensively investigated by chemical engineers (Shacham *et al.*, 1982; Westerberg *et al.*, 1979). Their concern is chiefly very large scale systems and the implementation of procedures capable of handling many hundreds of variables. As previously mentioned, the process control engineer is more often concerned with small or medium scale problems and his requirements are different: usually a computer library-based routine is adequate for solving eqn. 5.1 without resort to special flowsheeting programs.

One reason for a steady state simulation is to remove modelling errors. Steady state process measurements are often available whereas dynamic measurements may not be (Chapter 4). In that case the only check on the model validity is provided by the steady state simulation. Nor does linearisation render the steady state simulation superfluous, as the following example illustrates.

Example 5.2
We intend to model the nonlinear process

$$\mathring{x}(t) = \alpha x(t) + \beta u(t) + w \tag{5.2}$$

where $x(t)$, $u(t)$ and w are scalars, w a constant, by means of the nonlinear model

$$\mathring{z}(t) = \alpha z(t) + \beta u(t) + \hat{w} \tag{5.3}$$

The values of α, β and $\hat{w}$ are derived from theoretical considerations while $u(t)$ is measured and therefore known exactly. Obviously, the only difference between eqns. 5.2 and 5.3, assuming the same initial condition applies, is that $\hat{w} \neq w$. The model, eqn. 5.3, is now linearised, i.e.

$$\Delta\mathring{z}(t) = \alpha\Delta z(t) + \beta\Delta u(t) \tag{5.4}$$

where $\Delta z \triangleq z(t) - z$ and $\Delta u \triangleq u(t) - u$, and based upon the linear model a controller is designed. Let us say that the controller has the form

$$\Delta u(t) = f\Delta z(t) \tag{5.5a}$$

The use of state feedback in a controller will be justified in Chapter 6. To implement this controller we must supply

$$u(t) = f\Delta z(t) + u \tag{5.5b}$$

where u (and the corresponding steady-state z) is the point about which linearisation occurs. The behaviour of the closed loop is given by substituting eqn. 5.5*b* into eqn. 5.2:

$$\mathring{x}(t) \;=\; \alpha x(t) + \beta f \Delta z(t) + \beta u + w \qquad (5.6)$$

After some manipulation it is possible to express $\Delta z(t)$ in terms of $\Delta x(t)$, the deviation of $x(t)$ from some steady-state x. Thus

$$\Delta\mathring{x}(t) \;=\; (\alpha + \beta f)\Delta x(t) - (z - x)\beta f\, e^{\alpha t} + \alpha x + \beta u + w \qquad (5.7)$$

The term $(z - x)\beta f\, e^{\alpha t}$ is propagated by the inaccuracy in $\hat{w}$. It accounts for the incorrect dynamic behaviour which would result from implementing the controller. Note that it is the nonlinear model, eqn. 5.3 which is at fault and not the linear model, eqn. 5.4.

Pursuing this example further, we ask how such modelling inaccuracies could be detected and corrected. The answer is to perform a steady state simulation. The steady state corresponding to the nonlinear model, eqn. 5.3, is simulated (that is solved and displayed) and found to be given by the straight line of Fig. 5.2. Data from the process corresponding to various steady states are plotted in the same figure. Not only is the discrepancy obvious but in this example the remedy also.

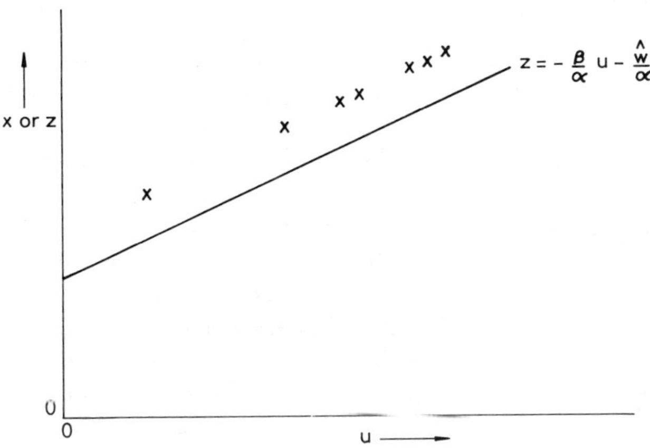

Fig. 5.2 *Steady-state simulation of Example 5.2*
× corresponds to one steady-state value measured from eqn. 5.2

We now return to the problem of how to find a solution to eqn. 5.1. There are many approaches (see, for example, Scheid, 1968) to this, the classical mathematical problem of finding the roots of one or more equations. For our purposes it is sufficient to describe just two, the first of which is preferable for smaller problems.

The first approach is to use numerical integration: eqn. 5.1 is converted to the differential equation

$$\mathring{x}(t) \;=\; f[x(t)] \qquad (5.8)$$

and this equation is numerically integrated from some initial guess $\tilde{x} = x_0$ satisfying $0 = f(\hat{x})$ until a steady state is reached. Assuming that eqn. 5.1 has a unique solution this is then equal to the steady state of eqn. 5.8. Various techniques for performing the numerical integration are described in the next sections of this chapter. This approach, which is suitable for simultaneous equations, is limited by the time required to perform the integration but is simple and straightforward to apply.

The second method is the Newton or Newton–Raphson procedure, described in most elementary mathematical texts. The Newton–Raphson procedure is extremely powerful having second-order convergence (Scheid, 1968). However, there are some drawbacks (Shacham *et al.*, 1982) which may necessitate the use of other techniques such as Quasi-Newton or Steepest Descent methods in very obstinate cases.

The Newton–Raphson procedure, as well as many other root finding algorithms are usually available in standard scientific computer software libraries.

5.4 Dynamic simulation

The type of model being used will give rise to simulations which are deterministic or stochastic, discrete or continuous in time. Many simulations will be deterministic and continuous, and our attention focuses on these in this section. One of the main reasons for simulating discrete-time models (for example the EDS) is the difficulty of calculating the state transition matrix (see Chapter 1). Stochastic simulations, and in particular Monte Carlo simulations, will play an ever increasing role in process control so this section contains some pertinent remarks; we start, however, with deterministic, continuous simulations.

The most widespread and successful approach to dynamic simulation is shown in Fig. 5.3. There are numerous numerical integration algorithms available (Fox, 1972; Johnson and Barney, 1977 and any scientific software library) and we describe here just one which has found wide application, the Runge–Kutta algorithm. In the next section of this chapter techniques are described which can out-perform the Runge–Kutta method when used on certain 'difficult' models (stiff systems).

The Runge–Kutta method is now described. To simplify the description it is assumed that the model consists of just one nonlinear equation in one variable $x(t)$:

$$\mathring{x}(t) = f[x(t), t] \tag{5.9}$$

Fig. 5.4 shows what happens in the Runge–Kutta algorithm per time step Δ. There are four calculations which take place.

1 An approximate value of $x\left(t_0 + \dfrac{\Delta}{2}\right)$ is calculated from

$$x_1\left(t_0 + \frac{\Delta}{2}\right) = x(t_0) + \mathring{x}(t_0)\,\frac{\Delta}{2} \qquad (5.10a)$$

and $\mathring{x}_1\left(t_0 + \frac{\Delta}{2}\right)$ is calculated from the model, eqn. 5.9.

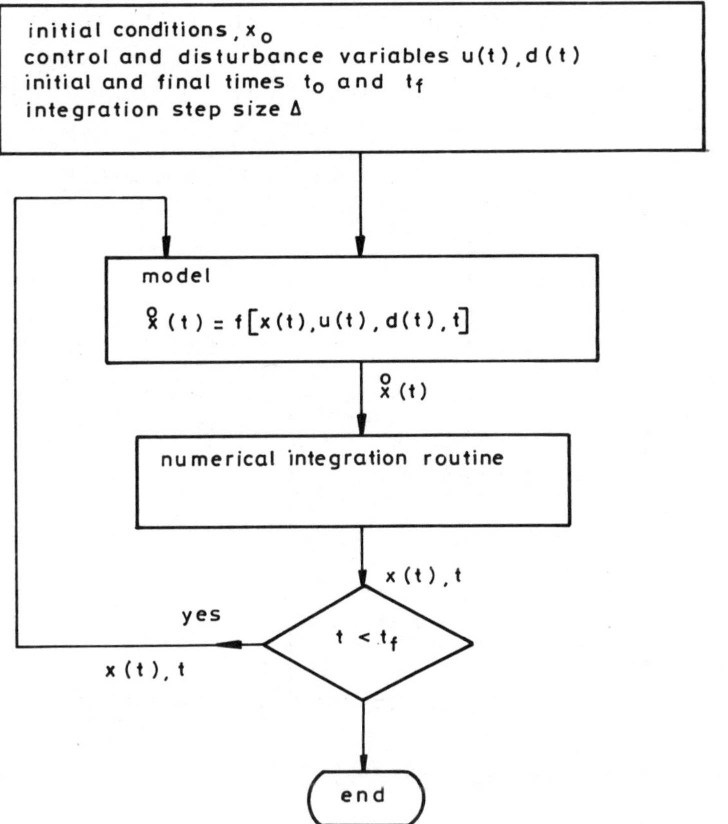

Fig. 5.3 *Dynamic simulation*

2 A second approximation to $x\left(t_0 + \frac{\Delta}{2}\right)$ is calculated from

$$x_2\left(t_0 + \frac{\Delta}{2}\right) = x(t_0) + x_1\left(t_0 + \frac{\Delta}{2}\right)\frac{\Delta}{2} \qquad (5.10b)$$

and again $\mathring{x}_2\left(t_0 + \frac{\Delta}{2}\right)$ is found from eqn. 5.9.

3 The first attempt at calculating $x(t_0 + \Delta)$ is now made from

$$x_3(t_0 + \Delta) = x(t_0) + \mathring{x}_2\left(t_0 + \frac{\Delta}{2}\right)\Delta \qquad (5.10c)$$

and the derivative $\mathring{x}_3\left(t_0 + \frac{\Delta}{2}\right)$ is evaluated.

4 Lastly, the best approximation to $x(t_0 + \Delta)$ is calculated using

$$x_4(t_0 + \Delta) = x(t_0) + \left[\mathring{x}(t_0) + 2\mathring{x}_1\left(t_0 + \frac{\Delta}{2}\right) + 2\mathring{x}_2\left(t_0 + \frac{\Delta}{2}\right) + \mathring{x}_3(t_0 + \Delta)\right]\frac{\Delta}{6} \qquad (5.11)$$

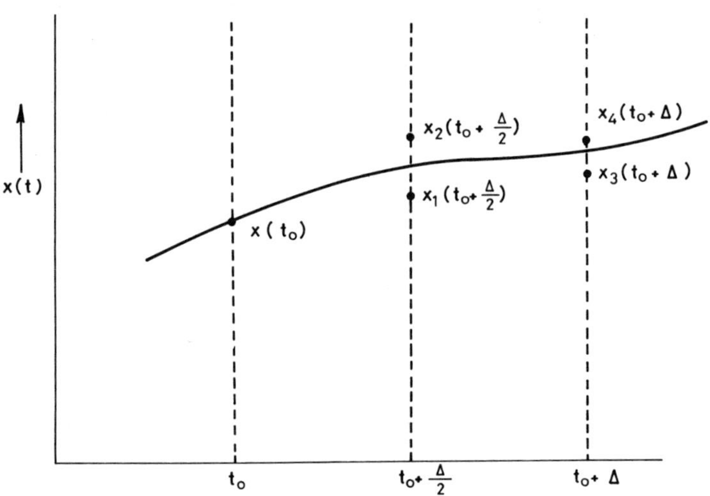

Fig. 5.4 *One time step during the Runge–Kutta integration*

The Runge–Kutta procedure is based upon a fourth-order Taylor series approximation, so that $x_4(t_0 + \Delta)$ will not be exactly equal to $x(t_0 + \Delta)$. However, no derivatives are required (as in the Taylor series) except the first derivative which is available from the model. For a derivation of the Runge–Kutta method and a discussion of the sources of error and convergence, texts on numerical analysis should be consulted (Scheid, 1968; Aggarwal, 1972). Note that the integration routine is self-starting and that numerous variants – including variable step length methods – are available (Johnson and Barney, 1977; Bui, 1981).

Lastly in this section, we turn to stochastic simulations. An assumption is made that a discrete-time model is involved (for example an EDS). Typically, additive white noise and an initial condition known only in terms of its first two moments are present. Monte Carlo simulations (Hammersley and Hanscombe, 1964) employ a sequence of random numbers from a random number generator to represent the stochastic process. Nearly every scientific software library has such a generator to produce discrete white noise (i.e. an uncorrelated sequence of zero-mean numbers). Such a sequence of random numbers may have either a uniform or a normal distribution and both mean and covariance may be set to the values desired. Exactly how this may be done is described elsewhere (Maisel and Gnugnoli, 1972; Bard, 1974).

Monte Carlo simulations require repeated runs: each run produces a discrete sample function of the model state and output. Using standard software library routines one can then evaluate the sample mean, covariance and probability distributions. If the model is ergodic then all its statistical properties can be determined from a single sample function or record; examples are stationary Gaussian Markov stochastic processes (Bendat and Piersol, 1971).

A test to check whether a discrete random process is white is not generally available in standard software libraries. Since in both simulation and in estimation and control such a test can be extremely useful we give one here, based upon the chi-square distribution (Stoica, 1977).

The sequence $\{w_k\}$ is white, with a risk of 5% if

$$\frac{N}{r_0^2}\sum_{i=1}^{\alpha} r_i^2 \leqslant \alpha + 1.95\sqrt{2\alpha} \tag{5.12}$$

Here, r_i is the sample covariance given by

$$r_i = \frac{1}{N}\sum_{K=0}^{N-i} w_K w_{K+i} \tag{5.13}$$

and α, N constants.

Details of other statistical tests can be found elsewhere (Gordon, 1969).

5.5 Stiff systems

Even for the small problems typically of interest to the process control engineer, simulations based on the Runge–Kutta algorithm described in the last section may be quite time consuming. The reason is that some states may evolve very quickly while others are much slower. Models where this happens are said to contain stiff differential equations.

For linear, time-invariant models a system is said to be stiff if

$$\frac{\max|\mathrm{Re}\{\lambda_i\}|}{\min|\mathrm{Re}\{\lambda_i\}|} \gg 1$$

where λ_i are the eigenvalues of the model.

Example 5.3

The following nonlinear model has been shown to adequately describe the dynamics of a fed-batch fermentation of baker's yeast:

$$\mathring{x}_1(t) = x_2(t)$$

$$\mathring{x}_2(t) = \frac{x_2^2(t)}{x_1(t)} - \alpha x_2(t) + \beta u(t)$$

$$y(t) = x_2(t)$$

Here $x_1(t)$ is the quantity of biomass (kg), $x_2(t)$ is rate of growth of biomass (kg/h), α and β experimentally determined constants, and finally $u(t)$ is the glucose feed rate in kg/h. To gain some insight into the dynamical behaviour of the fermentation, we linearise this nonlinear model about the reference trajectories

$$x_{1r}(t) = \gamma\ e^{\mu t}$$

$$x_{2r}(t) = \mu\gamma\ e^{\mu t}$$

$$u_r(t) = \frac{\alpha\mu\gamma\ e^{\mu t}}{\beta}$$

The linear model is then given by

$$\begin{bmatrix} \Delta\mathring{x}_1(t) \\ \Delta\mathring{x}_2(t) \end{bmatrix} = \begin{bmatrix} 0 & 1 \\ -\mu^2 & 2\mu - \alpha \end{bmatrix} \begin{bmatrix} \Delta x_i(t) \\ \Delta x_2(t) \end{bmatrix} + \begin{bmatrix} 0 \\ \beta \end{bmatrix} \Delta u(t)$$

$$\Delta y(t) = [0 \quad 1] \begin{bmatrix} \Delta x_1(t) \\ \Delta x_2(t) \end{bmatrix}$$

Taking $\alpha = 1.3\,h^{-1}$ and $\mu = 0.2\,h^{-1}$ we find that the two eigenvalues of the plant matrix (-0.853 and -0.047) differ by a factor of about 20. The process could be therefore termed a mildly stiff system.

For nonlinear, or even time-varying linear models the situation is much the same as with stability, where the eigenvalues *per se* tell us little. The usual approach to simulate stiff models is to refine or modify the Runge–Kutta routine already presented (Willoughby, 1974; Bui, 1981). Various so-called predictor–corrector algorithms (Scheid, 1968) and semi-implicit Runge–Kutta procedures such as the LSTIFF routine (Bui, 1981) have been developed, although perhaps the most widely used algorithm is the Gear routine (Gear, 1971). Texts on numerical analysis contain details of these types of routine.

Another approach, which at first seems to be completely different from the foregoing (although in fact is closely related to the conventional trapezoidal method of integration) is the block-pulse function method. This approach appears to have great potential for solving stiff systems (Palanisamy and Bhattacharya, 1982) and is applicable to nonlinear and time-varying models.

References

AGGARWAL, J. K. (1972): 'Notes on nonlinear systems' (Van Nostrand Reinhold)

BARD, Y. (1974): 'Nonlinear parameter estimation' (Academic Press)

BENDAT, J. S. and PIERSOL, A. G. (1971): 'Random data: analysis and measurement procedures' (Wiley)

BUI, T. D. (1981): 'Solving stiff differential equations in the simulation of physical systems', *Simulation*, **37**, 2, pp. 37–46

DAVISON, E. J. (1973): 'An algorithm for the computer simulation of very large dynamic systems', *Automatica*, **9**, pp. 665–675

ENRIGHT, W. (1979): 'On the efficient and reliable numerical solution of large linear systems of ODE's', *IEEE Trans. Auto. Control*, **AC-24**, pp. 905–908

FOX, P. (1972): 'A comparative study of computer programs for integrating differential equations', *Comm. ACM*, **15**, 11, pp. 941–948

GEAR, C. W. (1971): 'Numerical initial value problems in ordinary differential equations' (Prentice-Hall)

GORDON, G. (1969): 'Systems simulation' (Prentice-Hall)

HAMMERSLEY, J. M. and HANSCOMB, D. C. (1964): 'Monte Carlo methods' (Methuen)

JOHNSON, A. J. and BARNEY, J. E. F. (1977): 'Numerical methods for differential systems', eds LAPIDUS, L. and SCHIESSER, W. E. (Academic Press)

MAISEL, H. and GNUGNOLI, G. (1972): 'Simulation of discrete stochastic systems' (Science Research Associates)

PALANISAMY, K. R. and BHATTACHARYA, D. K. (1982): 'Analysis of stiff systems via single step method of block pulse functions', *Int. J. Sys. Sci.*, **13**, 9, pp. 961–968

SANNUTI, P. (1977): 'Analysis and synthesis of dynamic systems via block-pulse functions', *Proc. IEE*, **124**, 6, pp. 569–571

SCHEID, F. (1968): 'Numerical analysis' (McGraw-Hill)

SHACHAM, M., MACCHIETTO, S., STUTZMAN, F. and BABCOCK, P. (1982): 'Equation oriented approach to process flowsheeting', *Computers and Chem. Eng.*, **6**, 2, pp. 79–95

STOICA, P. (1977): 'A test for whiteness', *IEEE Trans. Auto. Control*, **AC-22**, 6, pp. 992–993

WESTERBERG, A. W., HUTCHINSON, H. P., MOTARD, R. L. and WINTER, P. (1979): 'Process flowsheeting' (Cambridge University Press)

WILLOUGHBY, R. A. (1974): 'Stiff differential equations' (Plenum)

Chapter 6

Process Control

6.1 Introduction

In an industrial context, the need for optimisation arises both before and after the commissioning of new plant. It is possible to distinguish between three problems requiring an optimal solution (see Fig. 6.1) although there is some overlap. The first and most important problem is the determination of the production capacity, run-time and starting date of the run, such that subject to the constraints of labour costs, down-time etc. profit is maximised. Next, each of the processes involved in the production must be operated optimally, that is to say that the most economic values must be selected for those variables of each process not specified during the production planning optimisation. Often the model of the process used for production planning is much more simplified than that used for process operation optimisation, and the cost criteria will be different. For new processes in the design stage, most of the equipment sizes such as pump capacities, stirrer-blade size, pipe diameters etc. will be known as a result of the process operation optimisation. Thirdly and last of all, deviations of the most important process variables from their optimal values (their 'set-points') must be minimised. Of course, the deviations are the result of disturbances acting on the process. This last optimisation problem – for so it can be formulated – will be called the setpoint control problem.

It is the purpose of this chapter to address these optimisation problems, and to thereby show when a deterministic or stochastic, static or dynamic, discrete-time or continuous-time process model is appropriate. As in Chapter 4 detailed proofs and historical details have been omitted. Furthermore, control system backup design and alarm system design are topics outside the scope of this text. The underlying philosophy of this chapter is similar to that first described by Athans (Athans, 1972).

Because of the similarities between production planning optimisation and process operation optimisation it is convenient to remove the distinction between them and talk henceforth of process optimisation. The similarities include the use of a deterministic, continuous-time (usually steady-state) process model and equations which describe physical or economical constraints on the process variables. In Section 6.2 process optimisation is discussed and some useful techniques for solving the most commonly encountered optimisation problems are presented.

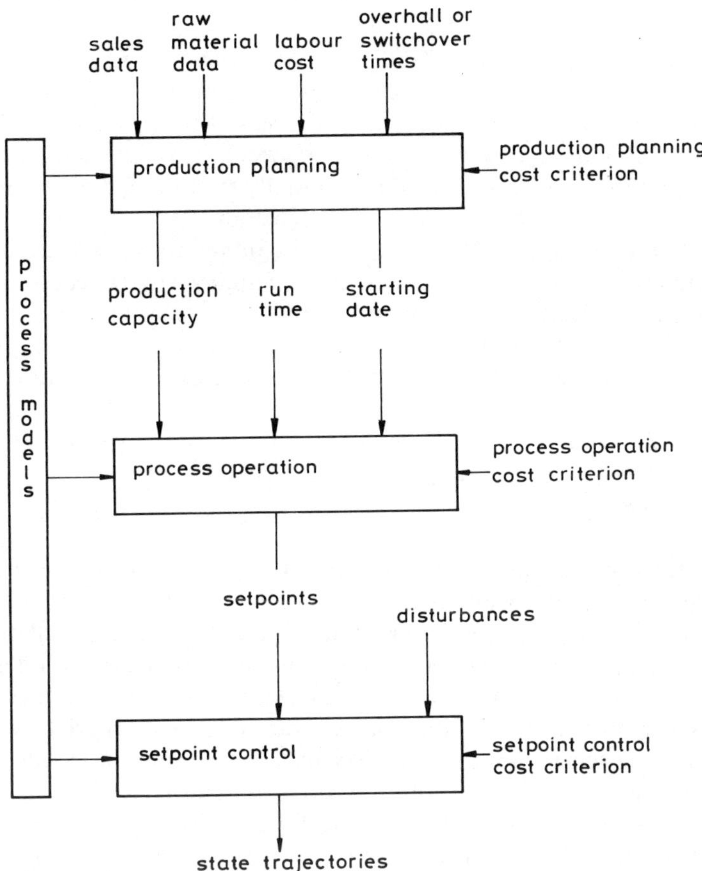

Fig. 6.1

The setpoint control problem, on the other hand, is an optimisation problem without constraints. This is because deviations of the controlled process variables from their set (i.e. optimal) values will be small if the controller is doing its job and so the variables are not likely (in a well-designed process) to be at their limiting values. Moreover, the setpoint control problem is dynamic, discrete-time (for online implementation) and essentially stochastic. As a prelude, however, we consider first the simpler deterministic version of the same problem in Section 6.3. This leads on to the solution of the setpoint control problem called the stochastic sampled-data linear optimal regulator (the LQG controller, as it is often called) in Section 6.4. The reasons for choosing the LQG controller in preference to other controllers which are available, such as the PID controller, are well documented elsewhere (Athans, 1972; Grimble, 1983) and will not be repeated here. Practical aspects of implementing the LQG controller, such as the inaccessibility of some state variables and the use of feedforward as well as feedback control, are discussed in Section 6.5.

Additional demands are often placed on the setpoint controller just described. For example, the operation of a process may be facilitated if there is little or no interaction between the state variables of the process. In Section 6.6 one type of optimal controller is described which can meet one of the commonest requirements of process control. Finally, Section 6.7 explores the difficult problem of actuator (and/or sensor) selection.

The last three sections of this chapter all contain considerable amounts of new material for the process control engineer. Supplementary reading is recommended to do justice to the considerable knowledge concerning optimisation and optimal controllers already existing. For the process control engineer two texts (Anderson and Moore, 1971; Kwakernaak and Sivan, 1972) are especially useful.

6.2 Process optimisation

In many industrial situations optimisation of the process variables (i.e. states and controls) is of primary importance. Because so many variables are involved – for example production planning should take into account all the many processes in the production chain – and because each of these variables may be subject to physical or economical constraints, only static optimisation is usually feasible. Static optimisation is the minimisation of a cost criterion, which is a function of the process variables, subject to a steady-state process model and constraints. The result of static optimisation is that all the process variables are given optimal constant, or fixed, values. In the context of process optimisation of many variables it is expedient to consider only deterministic optimisation since this leads to less computation. Even so, the computational burden is such that process optimisation invariably occurs offline.

An important exception to the static optimisation rule occurs when one or more processes are operated in a batch or fed-batch manner. Here, the optimisation problem is essentially a dynamic optimisation problem, where a functional (function of a function) of the process variables subject to a dynamic process model and constraints must be minimised. The result of dynamic optimisation is the optimal trajectories that the process variables must follow during the specified run time. Various numerical methods are available (Sage and White, 1977) to perform the dynamic optimisation, but these will not be described here. Indeed, the gradient method to be described shortly for static optimisation can easily be formulated to perform dynamic optimisation. Note that the problem of ensuring that the process variables follow their prescribed trajectories – the tracking problem – can be reduced to, and solved by, the same approach as used in solving the setpoint control problem (Kwakernaak and Sivan, 1972).

Most static optimisation problems cannot be solved by analytical methods (an important exception is noted later) so numerical search techniques (Beveridge and Schlechter, 1970; Sage and White, 1977) are employed. These techniques seek successively better feasible values for the process variables and are applied iteratively until no further improvement in reducing the costs is found, or until some predetermined degree of accuracy has been achieved. Techniques are characterised by the method of selecting a search direction and the distance of movement. If the gradient of the cost criterion is used to determine the search direction then the numerical optimisation technique is called a gradient technique. Gradient techniques are extremely powerful. However, in process optimisation it is usually too time-consuming to provide analytical expressions for the gradients required, while numerical gradient evaluation requires a considerable extra investment in programming (although the payoff may be justified if optimisation problems must be frequently solved). So here only a nongradient (or direct search) method will be presented, called the *Complex Method* (Box, 1965).

The problem of interest in static process optimisation is to minimise a nonlinear (scalar) cost criterion

$$J(u) \quad = \quad \theta(x, u) \tag{6.1}$$

subject to the steady-state (nonlinear) process model

$$0 \quad = \quad f(x, u) \tag{6.2}$$

Apart from satisfying the requirement that $\theta(x, u) \geqslant 0$ for all x and u, the process control engineer must endeavour to endow the mathematics with as much information as possible concerning the optimisation problem at hand through the choice of the cost criterion (Beveridge and Schechter, 1970).

In addition, inequality constraints

$$a \leqslant g(x, u) \leqslant b \tag{6.3}$$

that denote physical or economical restrictions applying to the process variables must be accounted for. Note that it is sometimes possible to infer restrictions on individual process variables (that then reduce the space in which the search for the optimal point need be conducted) from the inequality constraints of eqn. 6.3.

Before describing the Complex Method in detail a geometric interpretation of this approach to optimisation will be given (Beveridge and Schechter, 1970). A number, k, of feasible points, or vertices, are chosen to adequately cover the $(n + m)$ dimensional search space. Corresponding to each vertex is a cost, evaluated from eqn. 6.1. The vertex yielding the greatest cost is rejected and replaced by a point in space lying α ($\alpha \geqslant 1$) times as far from the centroid of the remaining vertices (i.e. their average value) as the distance of this rejected vertex but in a direction defined by a vector pointing from the rejected point to the centroid, see Fig. 6.2.

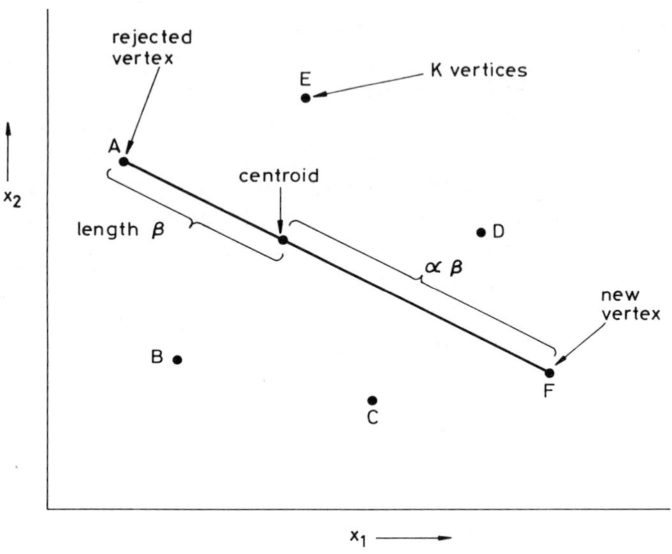

Fig. 6.2 *Complex method for two variables*

If either the cost corresponding to the new vertex is still the greatest or the new vertex violates a constraint of the problem, the vertex is moved halfway in towards the centroid. This movement is repeated until a valid point is obtained. Provided the search space is convex, the vertices eventually contract into the centroid, which is the optimum and the solution of the problem. For nonconvex space the method may be modified (Guin, 1968).

An algorithm to perform the Complex Method contains the following steps.

Initialisation stage:

1 Choose $m + 1 < k \leqslant 2m$ feasible values of the vector u, i.e. $\{u_1, u_2, \ldots, u_k\}$.
2 Calculate from eqn. 6.2 the corresponding vector set $\{x_1, x_2, \ldots, x_k\}$.
3 Test feasibility using eqn. 6.3. If any constraint is violated, go to step 1.

Iterative stage:

4 Calculate $J(u_i)$ for $i = 1, 2, \ldots, k$. Find most expensive vertex (u_j, x_j) where $J(u_j) > J(u_i)$ for all i.

5 Compute new u_j from old u_j using

$$u_j = \frac{(1+\alpha)}{k} \sum_{\substack{i=0 \\ i \neq j}}^{k} u_i - \alpha u_j$$

where $\alpha \geqslant 1$.

6 Calculate from eqn. 6.2 x_j.

7 If $J(u_j) \geqslant J(u_i)$ for all i, or vertex (u_j, x_j) not feasible,

$$u_j = \frac{1}{2}\left(\frac{1}{k} \sum_{\substack{i=0 \\ i \neq j}}^{k} u_i + u_j\right)$$

and go to step 6.

8 If all $J(u_i)$ same to within acceptable tolerance stop. Otherwise go to step 5.

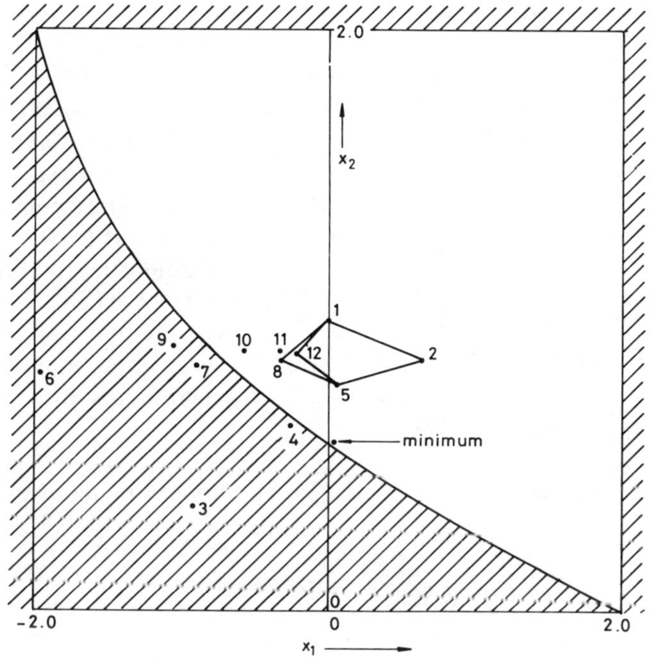

Fig. 6.3 *Complex method of Example 6.1*
▨ unfeasible region (restriction violated)

The value chosen for α is not critical; usually $\alpha = 1.3$ is taken. The above algorithm may be initialised by one of many techniques (see, for example, Beveridge and Schechter, 1970) and its use is demonstrated in the following example.

Example 6.1
Minimise the cost criterion

$$J(x_1, x_2) = 10x_1^2 + x_2^2$$

subject to the restriction that

$$x_1 + 2 - (x_2 - 2)^2 \geqslant 0$$

Table 6.1

Step no.	x_1	x_2	$J(x_1, x_2)$	remarks
1	0	1	1.0	chosen start point
2	0.647	0.872	4.95	randomly chosen x_1, x_2
3	−0.843	0.36	*	centroid (0.324, 0.936) generate new trial pair
4	−0.260	0.65	*	centroid as above generate new trial pair
5	0.032	0.793	0.639	
6	−1.993	0.822	*	centroid (0.266, 0.888) generate new trial pair
7	−0.884	0.855	*	generate new trial pair
8	−0.329	0.872	1.840	4 vertex figure complete worst point (0.647, 0.872) centroid other points is (−0.099, 0.888)
9	−1.069	0.910	*	generate new trial pair
10	−0.584	0.899	4.216	this is worst point of all 4: generate new trial pair
11	−0.341	0.894	1.964	
12	−0.2201	0.891	1.278	choose this point to replace vertex (0.647, 0.872)
.				
.				
.				
.				
.				
.				
.				
62	0.021	0.578	0.339	algorithm stops due to insignificant change in cost

* x_1, x_2 violate restriction

and that

$$-2 \leqslant x_1 \leqslant 2$$

$$0 \leqslant x_2 \leqslant 2$$

We choose a four vertex figure to cover the search area. The progress of the Complex algorithm is shown in Fig. 6.3, while the first few calculations are shown in Table 6.1.

Finally, we mention one very special case of static optimisation which permits a closed-form of solution. Suppose the cost criterion is quadratic, i.e.

$$J(u) \quad = \quad x^{\mathrm{T}}Qx + u^{\mathrm{T}}Ru \tag{6.4}$$

where Q and R are symmetric positive definite matrices, and the process model is linear in x and u:

$$0 \quad = \quad Ax + Bu + c \tag{6.5}$$

where c is an $(n \times 1)$ constant vector. Then the optimal value of u (Sage and White, 1977) is given by

$$u^0 \quad = \quad -R^{-1}B^{\mathrm{T}}(AQ^{-1}A^{\mathrm{T}} + BR^{-1}B^{\mathrm{T}})^{-1}c \tag{6.6}$$

6.3 Deterministic setpoint control

Having found the optimal values of all the process variables by the Complex Method or some other technique we turn our attention in this and the following two sections to the problem of keeping the variables at these 'setpoints'. In the beginning we adopt the rather naïve view that there are no disturbances affecting the process, and the problem of setpoint control is formulated as a deterministic dynamic optimisation problem. Later on more and more realism will be brought into the problem description.

The setpoint control problem has already been stated: it is the minimisation of the deviations caused by disturbances of process variables from their optimal values. Since the formulation of this problem as an optimal control problem requires all process variables to be included in the problem description the need must be stressed of reducing the process model order as far as possible, consistent with retaining the essential dynamical character of the process to be controlled (see Chapter 2). Later, in Section 6.6, we shall investigate what can be done to decompose the setpoint control problem if it is still unmanageably large, into smaller subproblems.

Furthermore, in this section we must assume that the actuators and sensors to be used have been selected. This selection is usually performed on the basis of the cost and installation costs of the equipment (but see Section 6.7).

Much valuable insight may be gained by introducing the subject of setpoint control in a deterministic setting. However, we should ask whether, without any disturbances, there is still a relevant and worthwhile problem to be solved. It turns out that there is. In the deterministic version of the setpoint control problem we attempt to annihilate an (initial) displacement of the state of the process from its optimal value. We assume that the cause of the displacement is unknown but that its magnitude (i.e. the initial state) can be measured. Fig. 6.4 indicates that since optimal values of both process states, denoted by x_{ref}, and control variables, denoted by u_{ref}, are available from the static optimisation of Section 6.2, it is possible to implement the setpoint controller in terms of perturbation variables, $x - x_{ref}$ etc., at least when all the process states are measurable. This restriction is removed in Section 6.5. Now if we choose to linearise the process model about x_{ref} and u_{ref} then the setpoint optimisation can proceed using the linear process model of Chapter 2. In fact since the setpoint controller will need to work online it is much more useful to base the setpoint optimisation upon the EDS of Chapter 3. Thereby both sampling and reconstruction of the signals are taken into account. We stress once again that the simplest possible EDS (see Chapter 2) should be used to describe the system dynamics.

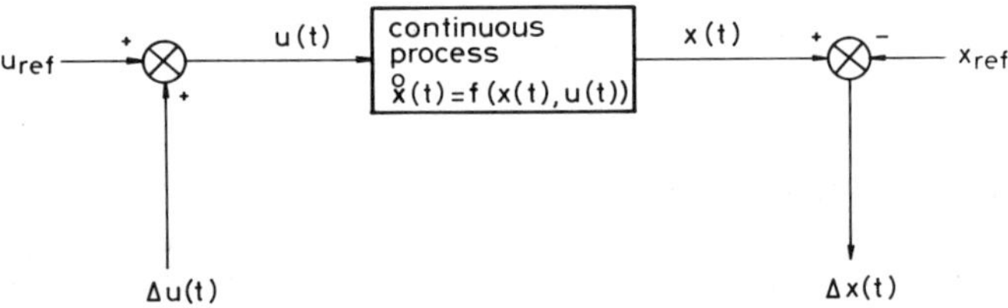

Fig. 6.4

It is logical to formulate the setpoint control problem with an integral cost criterion containing continuous-time variables. The latter ensures that deviations of the process variables from their setpoint (i.e. optimal) values at all times, not just at the times of sampling, can be accounted for. The criterion is to be minimised over an infinitely long interval of time since this is the most realistic assumption for process control. This also results in a simpler controller than would be obtained in the finite time-interval case. Note that when disturbances are included in the problem formulation (as in the next section, Section 6.4) a slightly different cost criterion must be used.

To solve the setpoint control problem with EDS and continuous-time cost criterion a transformation must be made into the equivalent discrete-time optimisation problem (Levis, Schlueter and Athans, 1971). This converts a constrained optimisation problem (since the control $u(t)$ is required to be piecewise-constant between sampling instants) into an unconstrained optimisation problem. We begin our development of the deterministic setpoint controller with the following result (Kalman, 1960; de Koning, 1982) concerning this unconstrained discrete-time problem.

Given the noise-free EDS

$$x_{k+1} = \Phi_k x_k + \Gamma_k u_k \qquad (6.7a)$$

$$x(0) = x_0, \text{ deterministic} \qquad (6.7b)$$

and all the states available for measurement. Denote the sequence of controls $\{u_0, u_1, \ldots\}$ by U. The cost criterion is

$$J(x_0, U) = \sum_{i=0}^{\infty} (x_i^T Q_i x_i + 2x_i^T M_i u_i + u_i^T R_i u_i) \qquad (6.8a)$$

where Q_i and R_i are symmetric, $(n \times n)$ and $(m \times m)$ matrices, and M_i is a $(n \times m)$ matrix.

Furthermore, it is assumed that

$$Q_i - M_i R_i^{-1} M_i^T \geqslant 0 \text{ for all } i \qquad (6.8b)$$

and

$$R_i > 0 \text{ for all } i \qquad (6.8c)$$

Then, under certain conditions, the unique U which minimises eqn. 6.8 subject to eqn. 6.7 exists and is given by

$$u_k^0 = -F_k x_k \qquad (6.9)$$

where

$$F_k = R_k^{-1} M_k^T + (R_k + \Gamma_k^T \Sigma_k \Gamma_k)^{-1} \Gamma_k^T \Sigma_k S_k \qquad (6.10a)$$

$$S_k = \Phi_k - \Gamma_k R_k^{-1} M_k^T \qquad (6.10b)$$

and Σ_k is the non-negative definite solution to the steady-state matrix Riccati equation:

$$\Sigma_k = Q_k - M_k R_k^{-1} M_k^T + S_k^T [\Sigma_k - \Sigma_k \Gamma_k (R_k + \Gamma_k^T \Sigma_k \Gamma_k)^{-1} \Gamma_k^T \Sigma_k] S_k \qquad (6.11)$$

Here, u_k^0 is called the *optimal control* and is generated by the *deterministic optimal discrete-time regulator*. The minimum cost is given by

$$J_{min}(x_0, U) = x_0^T \Sigma_k x_0 \qquad (6.12)$$

F_k is the $(m \times n)$ *feedback gain matrix*. The optimal regulator is given here in the form of linear feedback of all the states of the process model corresponding to the EDS of eqn. 6.7. The deterministic formulation of the problem in fact permits an alternative solution where only the initial state, x_0, and not the complete state trajectories are used in the optimal control u_k^0. However, in the more realistic stochastic version of the setpoint control problem which will be presented in the next section, state feedback is not only preferable but in fact essential.

The optimal regulator has a very similar structure to the steady-state Kalman filter presented in Chapter 4. This duality, as it is called, may be usefully exploited to enhance our knowledge of the regulator; it also means that the same computer software may be used for both Kalman filter and optimal regulator development (Johnson, 1983).

It has already been mentioned that the cost criterion of eqn. 6.8*a* is not really suitable for our purpose of setpoint control. Given the long sampling intervals typically encountered in process control, one would like to account for the behaviour of process variables between sampling instants as well as at the sampling instants; that is, use an integral criterion containing continuous-time variables. The following result meets this need.

Given the noise-free EDS

$$x_{k+1} = \Phi_k x_k + \Gamma_k u_k \tag{6.7a}$$

$$x(0) = x_0,\ \text{deterministic} \tag{6.7b}$$

and all the states available for measurement. The cost criterion is

$$J(x_0, U) = \int_0^\infty [x^T(t)Q(t)x(t) + u^T(t)R(t)u(t)]\,dt \tag{6.13}$$

where $Q(t)$ is a $(n \times n)$ symmetric, non-negative definite matrix and $R(t)$ is a $(m \times m)$ symmetric positive definite matrix, for all t.

Then, under certain conditions, the unique U which minimises eqn. 6.13 subject to eqn. 6.7 exists, and is given by

$$u_k^0 = -F_k x_k \tag{6.14}$$

where

$$F_k = \tilde{R}_k^{-1}\tilde{M}_k^T + (\tilde{R}_k + \Gamma_k^T\Sigma_k\Gamma_k)^{-1}\Gamma_k^T\Sigma_k\tilde{S}_k \tag{6.15}$$

and Σ_k is the non-negative definite solution to the steady-state matrix Riccati equation:

$$\Sigma_k = \tilde{Q}_k - \tilde{M}_k\tilde{R}_k^{-1}\tilde{M}_k^T + \tilde{S}_k^T[\Sigma_k - \Sigma_k\Gamma_k(\tilde{R}_k + \Gamma_k^T\Sigma_k\Gamma_k)^{-1}\Gamma_k^T\Sigma_k]\tilde{S}_k \tag{6.16}$$

Here, the optimal control u_k^0 is generated by the *deterministic optimal sampled-data regulator*. The minimum cost is given by eqn. 6.12, and the various matrices required to mechanise eqns. 6.15 and 6.16 are given by

$$\tilde{R}_k \triangleq \int_{kT_s}^{(k+1)T_s} [R(t) + \Gamma^T(t, kT_s)Q(t)\Gamma(t, kT_s)]\,dt \tag{6.17a}$$

$$\tilde{M}_k \triangleq \int_{kT_s}^{(k+1)T_s} \Phi^T(t, kT_s)Q(t)\Gamma(t, kT_s)\,dt \tag{6.17b}$$

$$\tilde{Q}_k \triangleq \int_{kT_s}^{(k+1)T_s} \Phi^T(t, kT_s)Q(t)\Phi(t, kT_s)\,dt \tag{6.17c}$$

$$\tilde{S}_k \triangleq \Phi_k - \Gamma_k \tilde{R}_k^{-1} \tilde{M}_k^T \tag{6.17d}$$

$$\Gamma_k = \Gamma[(k+1)T_s, kT_s] \tag{6.17e}$$

$$\Phi_k = \Phi[(k+1)T_s, kT_s] \tag{6.17f}$$

Proof:
The cost criterion of eqn. 6.13 can be expressed as

$$J(x_0, U) = \sum_{i=0}^{\infty} \int_{kT_s}^{(k+1)T_s} [x^T(t)Q(t)x(t) + u^T(t)R(t)u(t)]\,dt$$

Assuming a piecewise-constant control function, the state is described by

$$x(t) = \Phi(t, kT_s)x_k + \Gamma(t, kT_s)u_k$$

Hence the cost criterion becomes

$$\begin{aligned} J(x_0, U) &= \sum_{i=0}^{\infty} x_i^T \int_{kT_s}^{(k+1)T_s} \Phi^T(t, kT_s)Q(t)\Phi(t, kT_s)\,dt\; x_i \\ &\quad + 2x_i^T \int_{kT_s}^{(k+1)T_s} \Phi^T(t, kT_s)Q(t)\Gamma(t, kT_s)\,dt\; u_i \\ &\quad + u_i^T \int_{kT_s}^{(k+1)T_s} [R(t) + \Gamma^T(t, kT_s)Q(t)\Gamma(t, kT_s)]\,dt\; u_i \\ &= \sum_{i=0}^{\infty} x_i^T \tilde{Q}_i x_i + 2x_i^T \tilde{M}_i u_i + u_i^T \tilde{R}_i u_i \end{aligned}$$

It can be proved (Levis, Schlueter and Athans, 1971) that if $Q(t) \geqslant 0$ for all t, $\tilde{Q}_i - M_i R_i^{-1} M_i^T \geqslant 0$ for all i. Hence the problem has been transformed into an (unconstrained) equivalent discrete optimal control problem, and the result follows immediately from the discrete-time regulator.

If Φ_k, Γ_k, $Q(t)$ and $R(t)$ are constant matrices then the quantities required to calculate the feedback gain matrix simplify considerably (de Koning, 1980):

$$\tilde{R} = \int_0^{T_s} [R + \Gamma^T(t)Q\Gamma(t)]\,dt \simeq RT_s \tag{6.18a}$$

$$\tilde{M} = \int_0^{T_s} \Phi^T(t)Q\Gamma(t)\,dt \simeq \tfrac{1}{2}QBT_s^2 \tag{6.18b}$$

$$\tilde{Q} = \int_0^{T_s} \Phi^T(t)Q\Phi(t)\,dt \simeq QT_s + \tfrac{1}{2}(QA + A^TQ)T_s^2 \tag{6.18c}$$

$$\tilde{S} = \Phi - \Gamma\tilde{R}^{-1}\tilde{M}^T \simeq I + AT_s - \tfrac{1}{2}BR^{-1}B^TQT_s^2 \tag{6.18d}$$

$$\Gamma(T_s) = \int_0^{T_s} \Phi(t)B\,dt \simeq BT_s \tag{6.18e}$$

$$\Phi(T_s) = \exp(AT_s) \simeq I + AT_s \tag{6.18f}$$

The approximations given above are based upon the assumption that T_s is small.

The selection of the matrices $Q(t)$ and $R(t)$ are the responsibility of the process control engineer. Note that the cross-product term $\tilde{M}_k$ is a direct consequence of the signal sampling and reconstruction operations, and is fixed once $Q(t)$ has been chosen. Several methods exist to aid the selection of $Q(t)$ and $R(t)$; certain choices will result in increased closed-loop robustness (Wong, Stein and Athans, 1978; Grimble, 1983). However, more often than not the process control engineer must rely upon his intuition, backed up by simulations and rough guidelines (Athans, 1972; Kwakernaak and Sivan, 1972). A good start is to choose both weighting matrices to be time-invariant and diagonal. Then the larger $\|Q\|$ becomes, the greater the feedback gain and the faster deviations of the state from its optimal value are corrected. Conversely, the larger $\|R\|$ becomes, the smaller the feedback gain and the slower the system response. Sometimes it is possible to choose Q and R on economic grounds, but in setpoint control this is rarely the case; robustness – the invariance of the system response and stability to parametric changes in the process model – is a more useful criterion on which to base selection.

We turn now to the vital questions of existence and stability of the deterministic optimal sampled-data regulator. There is a duality with the results presented in Chapter 4 concerning the same questions about the Kalman Filter. The following holds (Anderson and Moore, 1981):

Assume that Φ_k, Γ_k, $Q(t)$ and $R(t)$ are all uniformly bounded and moreover that $R(t) \geqslant \alpha I > 0$. Then

1. If $(\tilde{S}_k, \Gamma_k)$ is uniformly stabilisable then Σ_k is bounded for all k.
2. If in addition $(\tilde{Q}_k^{1/2}, \tilde{S}_k)$ is uniformly detectable, then the optimally controlled closed-loop system is uniformly asymptotically stable.

The above conditions are sufficient (but not necessary) for the existence and stability of the optimal sampled-data regulator. In view of the definitions of detectability and stabilisability given in Chapter 3 it should be obvious that if the process to be controlled is uniformly asymptotically stable then the optimal regulator will exist and the closed-loop system

$$x_{k+1} = (\tilde{S}_k - \Gamma_k F_k)x_k \tag{6.19}$$

will be stable. Since closed-loop stability implies that $x_k \to 0$ as $k \to \infty$ we see that the optimal regulator fulfils the primary objective of nullifying the effect of an initial disturbance. It is clear from the presence of both $\tilde{Q}_k^{1/2}$ and $\tilde{S}_k$ in the conditions that sampling has an effect not only upon the stability of the closed-loop but can even determine whether an optimal regulator exists.

The optimal sampled-data regulator is a time-varying system. The following theorem (Kalman, 1960; Kwakernaak and Sivan, 1972) indicates when the feedback gain matrix, F_k, becomes constant.

In eqns. 6.7 and 6.13 let all the matrices be time invariant. Assume that $Q \geqslant 0$ and $R > 0$, with $(\tilde{Q}_k^{1/2}, \tilde{S})$ detectable and $(\tilde{S}, \Gamma)$ stabilisable. Then as $k \to \infty$, Σ_k converges to the unique, non-negative definite solution Σ of eqn. 6.16. The corresponding feedback gain matrix becomes the *steady-state feedback gain*, F.

Lastly in this section we look briefly at the robustness of the optimal sampled-data regulator to non-design conditions (Anderson and Moore, 1971; Safonov and Athans, 1977; Wong, Stein and Athans, 1978). Referring to Fig. 6.5, the tolerance of the closed-loop optimally controlled system to unmodelled non-linearities, perturbations etc. in the control channels is sought. In particular, we are looking for those values of N which destabilise the closed-loop system. The following result (Safonov and Athans, 1977) was derived for continuous-time systems so that some caution must be exercised when applying it to sampled-data systems such as those in which we are interested.

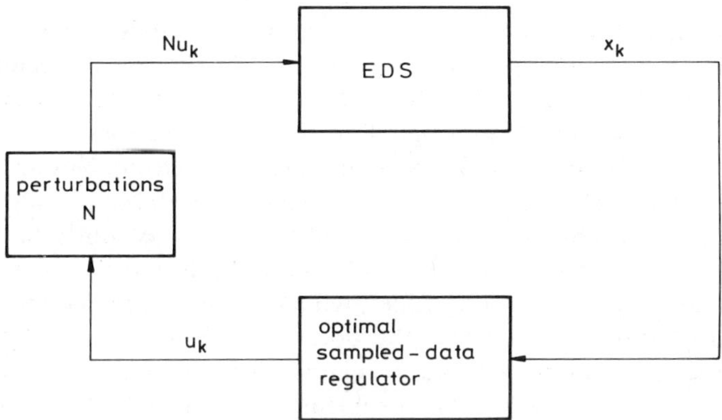

Fig. 6.5 *System robustness*

In eqns. 6.7 and 6.13 let all the matrices be time-invariant. Assume that $Q > 0$ and $R = \mathrm{diag}(r_i) > 0$, with $(\tilde{S}, \Gamma)$ stabilisable. Let one feedback loop $u^{(i)}$ be effected by an unmodelled, constant gain $n^{(i)} \neq 1$, where $i = 1, 2, \ldots, m$. Then as $T_s \to 0$, the closed-loop optimally regulated system remains asymptotically stable for any $n^{(i)} \geqslant 1/2$.

Note that because Q is assumed to be positive definite, $(Q^{1/2}, \tilde{S})$ is detectable. This result shows that the optimal sampled-data regulator can be expected to tolerate an infinite gain increase and at least 50% gain reduction in any one control channel. Provided that the conditions on Φ, Γ, Q and R stated above are met, the result holds for any process and any values of Q and R the process control engineer might choose. Actually, if some freedom exists (as it usually does) in the design of the regulator to permit Q and R to be chosen to enhance the robustness of the closed-loop, then we have (Wong, Stein and Athans, 1978):

With the same assumptions holding, if

$$N \triangleq \operatorname{diag}(n^{(i)}) > 1/2[I - (R^{1/2}FQ^{-1}F^{T}R^{1/2})^{-1}]$$

then as $T_s \to 0$ the closed-loop regulated system is asymptotically stable.

This result shows also which of the feedback loop gains can be reduced to zero (total loop failure situations) without destabilising the system. These robustness and integrity properties are among those which make the optimal regulator such a good choice of controller for process control.

6.4 Stochastic setpoint control

It is now the intention to increase the realism of the setpoint control problem discussed in the previous section by including disturbances. Recalling from Chapter 3 that the EDS modelled not only any disturbances acting on the process but also sampling and signal reconstruction errors and noise it is desirable to use the (stochastic) EDS in our design procedure. Not only does the inclusion of the full EDS add more realism to the problem but it results in a very practical controller. Only two problems will then still need our attention – the problem of unmeasured states, which is solved in Section 6.5, and the problem of the large dimension of the controller, which is resolved in Section 6.6. As in the previous section we state first the optimal discrete-time regulator before turning to the optimal sampled-data regulator. The stochastic optimal regulator problem (sometimes called the LQG problem) is formulated differently from the deterministic regulator problem of the last section; we have the following result (Meier, Larson and Tether, 1971; Kwakernaak and Sivan, 1972):

Given the EDS

$$x_{k+1} = \Phi_k x_k + \Gamma_k u_k + w_k \tag{6.20a}$$

$$E\{x_0\} = \bar{x}_0 \tag{6.20b}$$

$$E\{(x_0 - \bar{x}_0)(x_0 - \bar{x}_0)^{T}\} = P, \quad P \geqslant 0 \tag{6.20c}$$

$$E\{w_k\} = 0 \tag{6.20d}$$

$$E\{w_k w_l^{T}\} = W_k \delta_{kl}, \quad W_k \geqslant 0 \text{ for all } k \tag{6.20e}$$

$$E\{x_0 w_k^{T}\} = 0 \tag{6.20f}$$

and all the states may be measured. The cost criterion is

$$J(U) = \lim_{N\to\infty} \frac{1}{N} E\left\{ \sum_{i=0}^{N-1} x_i^T Q_i x_i + 2x_i^T M_i u_i + u_i^T R_i u_i \right\} \tag{6.21}$$

where $Q_k - M_k R_k^{-1} M_k^T$ is a symmetric, non-negative definite matrix for all k, and R_k is a symmetric positive definite matrix for all k.

Then, under certain conditions, the unique U which minimises eqn. 6.21 subject to eqn. 6.20 exists and is given by eqns. 6.9 to 6.11. The minimal cost is given by

$$J_{\min} = \operatorname{tr}(\Sigma_k W) \tag{6.22}$$

where $W_k \to W$ as $k \to \infty$.

The cost criterion has been formulated in a slightly different way to eqn. 6.8*a* so as to ensure that costs do not continue to increase as time progresses. It can be shown that the cost criterion may also be written as

$$J(U) = \lim_{N\to\infty} E\{x_N Q_N^T x_N + 2x_N^T M_N u_N + u_N^T R_N u_N\} \tag{6.23}$$

so that an alternative interpretation in terms of the ultimate cost per sampling instant is possible. Note from eqn. 6.22 that the possibility exists of zero costs if there are no disturbances (i.e. $W_k = 0$). However, as far as the optimal control u_k^0 which is the solution to the problem is concerned, the inclusion of disturbances has not made the optimal regulator more complicated; in fact, it has not changed it at all. If, as we have assumed, all the states are measured, and provided the actual value of the minimum cost is not required, then the disturbance statistics $\bar{x}_o$, P and W_k need not be known; the solution to the stochastic optimal regulator problem can proceed without them.

The foregoing problem is now modified to include the costs of deviations from the optimal static values of the states between sampling instants (Halyo and Caglayan, 1976; de Koning, 1984). This involves the use of an integral, rather than a sum criterion, as explained in the previous section. We have the following result.

Given the EDS

$$x_{k+1} = \Phi_k x_k + \Gamma_k u_k + w_k \tag{6.20a}$$

$$E\{x_0\} = \bar{x}_0 \tag{6.20b}$$

$$E\{(x_0 - \bar{x}_0)(x_0 - \bar{x}_0)^T\} = P, \quad P \geqslant 0 \tag{6.20c}$$

$$E\{w_k\} = 0 \tag{6.20d}$$

$$E\{w_k w_l^T\} = W_k \delta_{kl}, \quad W_k \geqslant 0 \text{ for all } k \tag{6.20e}$$

$$E\{x_0 w_k^T\} = 0 \tag{6.20f}$$

and all the states may be measured. The cost criterion is

$$J(U) = \lim_{t_f \to \infty} \frac{1}{t_f - t_0} E\left\{\int_{t_0}^{t_f} x^T(t)Q(t)x(t) + u^T(t)R(t)u(t)\, dt\right\} \quad (6.24)$$

where $Q(t)$ is a symmetric, non-negative definite matrix for all t, and $R(t)$ is a symmetric, positive definite matrix for all t.

Then, under certain conditions, the unique U which minimises eqn. 6.24 subject to eqn. 6.20 exists and is given by eqn. 6.14 to 6.16. The optimal control u_k^0 is generated by the *stochastic optimal sampled-data regulator*, and the minimal cost of using this regulator is given by

$$J_{\min} = \frac{1}{T_s}\left[\gamma + \lim_{k \to \infty} \frac{1}{k} \sum_{i=0}^{k-1} \operatorname{tr}(\Sigma_k W_k)\right] \quad (6.25)$$

where

$$\gamma = \int_{kT_s}^{(k+1)T_s} \operatorname{tr}[W(t)Q(t)]\, dt$$

with

$$W(t) = \int_{kT_s}^{kT_s + t} \Phi(\sigma, kT_s) W_k \Phi^T(\sigma, kT_s)\, d\sigma$$

The same sufficient detectability and stabilisability conditions ensure the existence and stabilisability of the stochastic optimal sampled-data regulator as the deterministic sampled-data regulator, with the proviso that W_k is a uniformly bounded matrix. If, as has been assumed, all the states of the process (and thus of the EDS) can be measured, the stochastic regulator is also just as robust as the deterministic regulator.

One important difference in the outcome of the deterministic and stochastic setpoint problems is that in the solution of the latter, *feedback* is essential. Of course, we have presented the solution of the deterministic setpoint problem in the form of a feedback control strategy, eqn. 6.14, but this was a somewhat arbitrary choice since the solution can also be written in the open-loop form (Kwakernaak and Sivan, 1972)

$$u_k^0 = G_k x_0 \quad (6.26)$$

When the more realistic, stochastic setpoint problem is solved it turns out that the feedback form of eqn. 6.14 is the only possible solution. The following example should make this assertion meaningful.

Example 6.2
Consider the scalar, time invariant EDS

$$x_{k+1} = \Phi x_k + \Gamma u_k + w_k$$

$$y_k = x_k$$

It is desired to annihilate the effect of any initial displacement x_0 given that w_k is discrete white noise with $E\{w_k\} = 0$ and $E\{w^2\}_k = w$. Both open-loop and closed-loop control are to be considered.

Application of the stochastic optimal sampled-data regulator gives a closed-loop system:

$$x_{k+1} = (\Phi - \Gamma f)x_k + w_k$$

assuming that f is the steady-state feedback gain. This stochastic system may be solved to give

$$\begin{aligned} x_k &= \tilde{\Phi}^k x_0 + \tilde{\Phi}^{k-1} w_0 + \tilde{\Phi}^{k-2} w_1 + \ldots + w_{k-1} \\ &= \tilde{\Phi}^k x_0 + \sum_{i=0}^{k-1} \tilde{\Phi}^{k-1-i} w_i \end{aligned}$$

where $\tilde{\Phi} \triangleq (\Phi - \Gamma f)$, so that using the given properties of the white noise

$$E\{x_k\} \triangleq \bar{x}_k = \tilde{\Phi}^k E\{x_0\} = \tilde{\Phi}^k \bar{x}_0 \tag{6.27}$$

$$E\{(x_k - \bar{x}_k)^2\} = \tilde{\Phi}^{2k} E\{(x_0 - \bar{x}_0)^2\} + w \sum_{i=0}^{k-1} \tilde{\Phi}^{2i} \tag{6.28}$$

Assuming the existence and stability of the stochastic optimal regulator, $\tilde{\Phi}^k \to 0$ as $k \to \infty$. Then from eqns. 6.27 and 6.28 it follows that both the mean and variance of the state decay away to zero as time progresses.

Now suppose that open-loop control was applied. The controlled stochastic system becomes

$$x_{k+1} = \Phi x_k + \Gamma g x_0 + w_k \tag{6.29}$$

assuming that g is the steady-state open-loop gain. Solving eqn. 6.29 gives

$$x_k = \Phi^k x_0 + \sum_{i=0}^{k-1} \Phi^{k-1-i} \Gamma g x_0 + \sum_{i=0}^{k-1} \Phi^{k-1-i} w_i$$

$$= \alpha_k x_0 + \sum_{i=0}^{k-1} \Phi^{k-1-i} w_i, \text{ say} \tag{6.30}$$

Once again using the given properties of white noise w_k we have that

$$E\{x_k\} = \bar{x}_k = \alpha_k \bar{x}_0 \tag{6.31}$$

$$E\{x_k - \bar{x}_k)^2\} = \alpha_k^2 E\{(x_0 - \bar{x}_0)^2\} + w \sum_{i=0}^{k-1} \Phi^{2i} \tag{6.32}$$

Superficially there is little changed. However, on examining the variance, eqn. 6.32, we see that if $|\Phi| > 1$ then the variance becomes unbounded as $k \to \infty$, *and we can do nothing about it*! Therefore, as this very simple example shows, open-loop control is useless for an unstable EDS; it is no wonder that the solution of the stochastic optimal sampled-data regulator problem therefore excludes this type of control strategy.

6.5 Regulator implementation

This section is devoted to three practical points which arise in the consideration of the stochastic optimal sampled-data regulator. First, we remove the restriction that all process states must be measurable, and discuss how both the closed-loop stability and its robustness is affected. Next the situation when one or more of the disturbances can be measured is considered; this introduces feedforward control to complement the (now familiar) feedback control. Lastly, we show how to take care of changing setpoints.

When not all of the process states can be measured it is not possible to apply the setpoint regulator of the last section. It can be proved (Joseph and Tou, 1961; de Koning, 1984) under suitable conditions that if the optimal control u_k^0 may be expressed in terms of $y_0, y_1, \ldots, y_{k-1}$ and $u_0, u_1, \ldots, u_{k-1}$ then the following result holds:

Given the EDS

$$x_{k+1} = \Phi_k x_k + \Gamma_k u_k + w_k \tag{6.33}$$

$$y_k = C_k x_k + v_k \tag{6.34}$$

where

$$E\{x_0\} = \bar{x}_0 \tag{6.35a}$$

$$E\{(x_0 - \bar{x}_0)(x_0 - \bar{x}_0)^{\mathrm{T}}\} = P, \quad P \geqslant 0 \tag{6.35b}$$

$$E\{w_k\} = E\{v_k\} = 0 \tag{6.35c}$$

$$E\{w_k w_l^{\mathrm{T}}\} = W_k \delta_{kl}, \quad W_k \geqslant 0 \tag{6.35d}$$

$$E\{w_k v_l^{\mathrm{T}}\} = S_k' \delta_{kl} \tag{6.35e}$$

$$E\{v_k v_l^{\mathrm{T}}\} = V_k \delta_{kl}, \quad V_k > 0 \tag{6.35f}$$

$$E\{x_0 w_k^{\mathrm{T}}\} = 0 \tag{6.35g}$$

$$E\{x_0 v_k^{\mathrm{T}}\} = 0 \tag{6.35h}$$

and the cost criterion

$$J(U) = \lim_{t_f \to \infty} \frac{1}{t_f - t_0} E\left\{\int_{t_0}^{t_f} [x^{\mathrm{T}}(t)Q(t)x(t) + u^{\mathrm{T}}(t)R(t)u(t)]\,\mathrm{d}t\right\} \tag{6.36}$$

where $Q(t) \geqslant 0$ and $R(t) > 0$ for all t. Then the unique sequence of controls which minimises eqn. 6.36 subject to eqns. 6.33 and 6.34 is, if it exists,

$$u_k^0 = -F_k \hat{x}_{k/k-1} \tag{6.37}$$

where F_k is the feedback gain matrix given by eqns. 6.14 to 6.16 and $\hat{x}_{k/k-1}$ is the linear minimum variance estimate given by the Kalman filter (eqns. 4.43 to 4.47) of x_k

---

Hence the optimal setpoint regulator with incomplete state measurements decomposes into two quite separate problems which can be solved independently: the calculation of the optimal feedback gain matrix F_k and the calculation of the Kalman filter. This result is known as the *separation principle*. Fig. 6.6 shows the implementation of the optimal regulator; it is as well to remark, perhaps superfluously, that it is clear from Fig. 6.6 that both the Kalman filter and the optimal regulator are based on an EDS using perturbation variables. u^0 and y^0 are constant values (the setpoints) derived from the process optimisation described in Section 6.2.

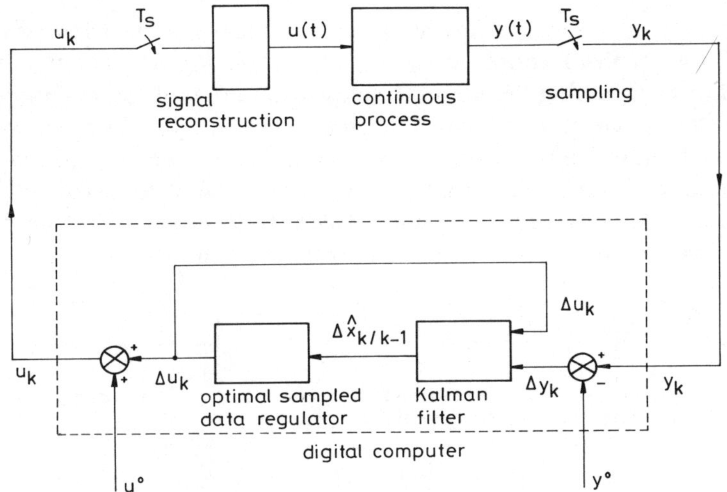

Fig. 6.6 *Optimal setpoint control*
Note: in this figure perturbation variables have been prefixed by Δ. In the text this is not always so

The inclusion of the Kalman filter in the closed-loop shown in Fig. 6.6 has an effect on the robustness of the controller. To examine this we must first consider the stability of the closed-loop system, see eqn. 6.19. To simplify matters, we restrict the discussion to those time-invariant processes which satisfy the conditions guaranteeing stability of the KF and the stability of the optimal sampled-data regulator. We would expect the closed-loop system, i.e.

$$\begin{bmatrix} x_{k+1} \\ \hat{x}_{k+1/k} \end{bmatrix} = \begin{bmatrix} S & -\Gamma F \\ KC & S - KC - \Gamma F \end{bmatrix} \begin{bmatrix} x_k \\ \hat{x}_{k/k-1} \end{bmatrix} + \begin{bmatrix} w_k \\ Kv_k \end{bmatrix} \tag{6.38}$$

also to be stable. Recall that the eigenvalues of a dynamical system are invariant under a system similarity transformation. Hence the eigenvalues of the closed loop matrix of eqn. 6.38 are the same as the eigenvalues of the matrices

$$\begin{bmatrix} I & 0 \\ -I & I \end{bmatrix} \begin{bmatrix} S & -\Gamma F \\ KC & S - KC - \Gamma F \end{bmatrix} \begin{bmatrix} I & 0 \\ I & I \end{bmatrix} = \begin{bmatrix} S - \Gamma F & -\Gamma F \\ 0 & S - KC \end{bmatrix} \tag{6.39a}$$

i.e.

$$\lambda_i[S - \Gamma F] \quad \text{and} \quad \lambda_i[S - KC] \tag{6.39b}$$

where the last result follows from applying Schur's formula for the determinant of a partitioned matrix (Chapter 1). The stability of the closed-loop is therefore assured in the idealised situation if both the KF and the optimal sampled-data regulator are stable.

To investigate robustness we once again consider unmodelled constant gains affecting the feedback loops, as in Fig. 6.5 of Section 6.3. Doyle (1978) has shown that by including the KF in the closed-loop the robustness properties of the optimal regulator, as described in Section 6.3, are lost. In other words, an unmodelled input gain of the process could cause the closed-loop to be destabilised. Coinsider the special case when just one of the feedback loops is attenuated by n, where n is some unmodelled constant gain. The more general situation we would wish to handle is shown in Fig. 6.7.

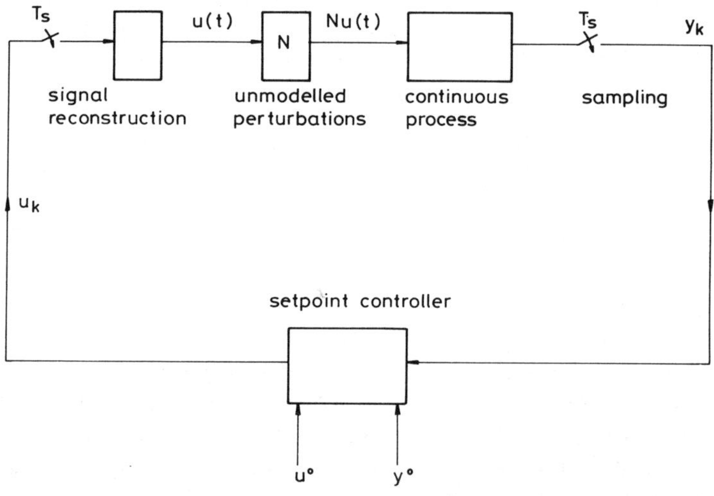

Fig. 6.7 *Robustness of the regulator and Kalman filter*

The dynamics of the closed-loop system is described by the equation

$$\begin{bmatrix} x_{k+1} \\ \hat{x}_{k+1/k} \end{bmatrix} = \begin{bmatrix} S & -\Gamma NF \\ KC & S - KC - \Gamma F \end{bmatrix} \begin{bmatrix} x_k \\ \hat{x}_{k/k-1} \end{bmatrix} + \begin{bmatrix} w_k \\ Kv_k \end{bmatrix} \tag{6.40}$$

A system similarity transformation applied to eqn. 6.40:

$$\begin{bmatrix} I & -I \\ 0 & I \end{bmatrix} \begin{bmatrix} S & -\Gamma NF \\ KC & S - KC - \Gamma F \end{bmatrix} \begin{bmatrix} I & I \\ 0 & I \end{bmatrix}$$

$$= \begin{bmatrix} S - KC & \Gamma(I - N)F \\ KC & S - \Gamma F \end{bmatrix} = \begin{bmatrix} S - KC & -ng_i f_i^{\mathrm{T}} \\ KC & S - \Gamma F \end{bmatrix} \tag{6.41}$$

where g_i and f_i^{T} are the ith column and row of Γ and F, respectively, maintains the closed-loop eigenvalues and provides some insight into the robustness problem. If KC is zero, for example, the closed loop eigenvalues are the eigenvalues of S and those of $(S - \Gamma F)$. Note that the condition that $KC = 0$ is not equivalent to assuming complete state measurement. For the important case of an open-loop stable EDS the foregoing indicates that to enhance the robustness of the closed-loop system, the KF should be designed such that KC is as small as possible (i.e. near zero elements). The effect of n on the closed-loop eigenvalues may also be studied using Gershgorin's Theorem (Rosenbrock and Storey, 1970), as in the following example (Doyle, 1978).

Example 6.3

The process to be controlled is described by the following mathematical model:

$$\mathring{x}(t) = \begin{bmatrix} 1 & 1 \\ 0 & 1 \end{bmatrix} x(t) + \begin{bmatrix} 0 \\ 1 \end{bmatrix} u(t) + n(t) \tag{6.42a}$$

$$y(t) = [1 \quad 0]x(t) + v(t) \tag{6.42b}$$

where $n(t)$ and $v(t)$ are zero-mean white noises with

$$E\{n(t)n^{\mathrm{T}}(s)\} = \sigma \begin{bmatrix} 1 & 1 \\ 1 & 1 \end{bmatrix} \delta(t - s), \quad \sigma > 0 \tag{6.43a}$$

$$E\{v(t)v^{\mathrm{T}}(s)\} = \delta(t - s) \tag{6.43b}$$

The EDS (see Chapter 3) of this process may be shown to be given for small T_s approximately by

$$x_{k+1} = \begin{bmatrix} 1 + T_s & T_s \\ 0 & 1 + T_s \end{bmatrix} x_k + \begin{bmatrix} 0 \\ T_s \end{bmatrix} u_k + w_k \tag{6.44a}$$

$$y_k = [1 \quad 0]x_k + v_k \tag{6.44b}$$

$$E\{w_k w_l^{\mathrm{T}}\} = \sigma T_s \begin{bmatrix} 1 & 1 \\ 1 & 1 \end{bmatrix} \delta_{kl} \tag{6.45a}$$

$$E\{v_k v_l^{\mathrm{T}}\} = \gamma \delta_{kl} \tag{6.45b}$$

where γ is some very large positive number. The (continuous-time) cost criterion is given by eqn. 6.20 with (say)

$$Q = \alpha\begin{bmatrix}1 & 1\\ 1 & 1\end{bmatrix}, \quad \alpha > 0 \tag{6.46a}$$

$$R = 1 \tag{6.46b}$$

Using this information, the various parameters needed to compute the stochastic optimal setpoint controller may be evaluated from eqn. 6.14. We make the assumption here that T_s is so small that any terms in T_s^2 or higher powers of T_s may be neglected. Then from eqn. 6.14

$$\tilde{R} \simeq RT_s = T_s \tag{6.47a}$$

$$\tilde{M} \simeq \frac{T_s^2}{2}QB = \begin{bmatrix}0\\ 0\end{bmatrix} \tag{6.47b}$$

$$\tilde{Q} \simeq QT_s + \frac{T_s^2}{2}(QA + A^TQ) = \alpha T_s\begin{bmatrix}1 & 1\\ 1 & 1\end{bmatrix} \tag{6.47c}$$

$$S \simeq I + AT_s - \frac{T_s^2}{2}BR^{-1}B^TQ = \begin{bmatrix}1 + T_s & T_s\\ 0 & 1 + T_s\end{bmatrix} \tag{6.47d}$$

$$\Gamma \simeq BT_s = \begin{bmatrix}0\\ T_s\end{bmatrix} \tag{6.47e}$$

$$Q \simeq I + AT_s = \begin{bmatrix}1 + T_s & T_s\\ 0 & 1 + T_s\end{bmatrix} \tag{6.47f}$$

After some calculations the optimal feedback gain matrix is found from eqns. 6.10 and 6.11 to be

$$F = [2 + \sqrt{(4 + \alpha)}][1 + T_s \quad 1 + 2T_s] \tag{6.48}$$

$$\triangleq [\eta_1 \quad \eta_2] \tag{6.49}$$

Using again the assumption that second and higher order powers of T_s may be neglected, the Kalman filter gain is found from eqn. 4.45 to be

$$K = \frac{\beta}{\gamma + \beta}\begin{bmatrix}1 + 2T_s\\ 1 + T_s\end{bmatrix} \triangleq \begin{bmatrix}\kappa_1\\ \kappa_2\end{bmatrix} \tag{6.50}$$

$$\beta = \left(2\gamma + \frac{\sigma}{2}\right)T_s + \{\sigma\gamma T_s\}^{\frac{1}{2}}$$

The closed-loop system is given by eqn. 6.40, and on substituting eqns. 6.49 and

6.50 we find that

$$\begin{bmatrix} S - KC & -ng_i f_i^{\mathrm{T}} \\ KC & S - \Gamma F \end{bmatrix}$$

$$= \begin{bmatrix} 1 + T_s - \kappa_1 & T_s & 0 & 0 \\ -\kappa_2 & 1 + T_s & -nT_s\eta_1 & -nT_s\eta_2 \\ \kappa_1 & 0 & 1 + T_s & T_s \\ \kappa_2 & 0 & -T_s\eta_1 & -T_s\eta_2 \end{bmatrix} \tag{6.52}$$

It is straightforward to show that the eigenvalues of the above matrix are the same as the eigenvalues of the matrix

$$\begin{bmatrix} * & * & * & * \\ -\kappa_2 & 1 - T_s & 0 & nT_s\eta_2 \\ * & * & * & * \\ * & * & * & * \end{bmatrix} \tag{6.53}$$

where the actual values of the elements other than those of the second row do not concern us, except that they are not functions of n, the unmodelled perturbation. Applying Gershgorin's theorem four discs encircling the region in the complex plane wherein the closed-loop eigenvalues lie may be constructed. Only one disc, centred at the point $(1 - T_s)$ and with a radius of $|\kappa_2| + |nT_s\eta_2|$ is dependent upon n. When $n = 1$, i.e. there are no unmodelled perturbations, the discs are such that all the closed-loop eigenvalues lie within the unit circle centred at the origin of the complex plane. When $\kappa_2 = \kappa_1 = 0$ the same situation occurs.

We consider next the situation when one or more of the disturbances acting on the process can be measured. Let these measurable disturbances be collected together into the $(\alpha \times 1)$ dimensional vector d_k and let their dynamics (after sampling) be described by the model, see Fig. 6.8,

$$d_{k+1} = H_k d_k + n_k \tag{6.54}$$

where n_k is discrete white noise having zero-mean. The remaining mathematical description of the situation is exactly that of eqns. 6.33 to 6.35 where now w_k' are those unmeasurable disturbances acting on the process.

Augmenting the state vector, x_k, with d_k we may write eqns. 6.33 and 6.54 as

$$\begin{bmatrix} x_{k+1} \\ d_{k+1} \end{bmatrix} = \begin{bmatrix} \Phi_k & D_k \\ 0 & H_k \end{bmatrix} \begin{bmatrix} x_k \\ d_k \end{bmatrix} + \begin{bmatrix} \Gamma_k \\ 0 \end{bmatrix} u_k + w_k' \tag{6.55}$$

and the cost criterion, eqn. 6.36, is appropriately modified so that the state weighting matrix now has the form

$$\begin{bmatrix} Q(t) & 0 \\ 0 & 0 \end{bmatrix} \qquad (6.56)$$

This modified optimal sampled-data regulator problem can be solved using the same results as quoted in the previous two sections. After some manipulations we find that the optimal control is given by

$$u_k^0 = -F_k x_k - G_k d_k \qquad (6.57)$$

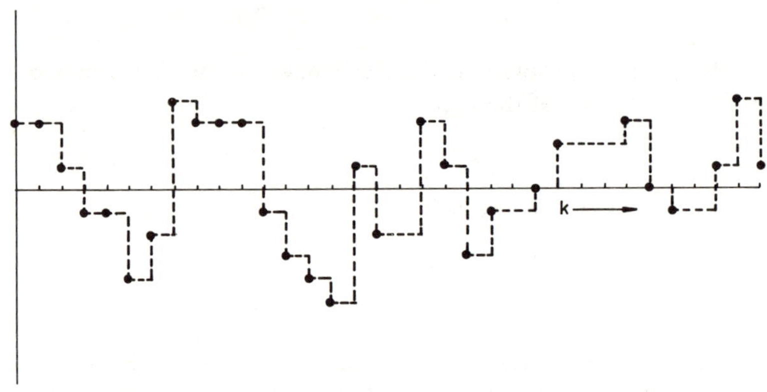

Fig. 6.8 *The 'random walk' of* $d_{k+1} = h_k d_k + n_k$

where F_k is given, as when $d_k = 0$, by eqns. 6.15 and 6.16, while

$$G_k = \tilde{R}_k^{-1}\bar{M}_k^{\mathrm{T}} + (\tilde{R}_k + \Gamma_k^{\mathrm{T}}\Sigma_k\Gamma_k)^{-1}\Gamma_k^{\mathrm{T}}(\Sigma_k\bar{S}_k + P_k H_k) \qquad (6.58)$$

Here, Σ_k is given by eqn. 6.16, P_k is the solution of the equation

$$\begin{aligned} P_k = {} & S_k^{\mathrm{T}}(\Sigma_k - \Sigma_k\Gamma_k(\tilde{R}_k + \Gamma_k^{\mathrm{T}}\Sigma_k\Gamma_k)^{-1}\Gamma_k^{\mathrm{T}}\Sigma_k)\bar{S}_k \\ & + S_k^{\mathrm{T}}(P_k - \Sigma_k\Gamma_k(\tilde{R}_k + \Gamma_k^{\mathrm{T}}\Sigma_k\Gamma_k)^{-1}\Gamma_k^{\mathrm{T}}P_k)H_k \\ & + \bar{N}_k - \bar{M}_k\tilde{R}_k^{-1}\bar{M}_k^{\mathrm{T}} \end{aligned} \qquad (6.59)$$

and the various newly-defined matrices are

$$\bar{M}_k \triangleq \int_{kT_s}^{(k+1)T_s} D^{\mathrm{T}}(t, kT_s)Q(t)\Gamma(t, kT_s)\,\mathrm{d}t \qquad (6.60a)$$

$$\bar{S}_k \triangleq D_k - \Gamma_k\tilde{R}_k^{-1}\bar{M}_k^{\mathrm{T}} \qquad (6.60b)$$

$$\bar{N}_k \triangleq \int_{kT_s}^{(k+1)T_s} \Phi^{\mathrm{T}}(t, kT_s)Q(t)D(t, kT_s)\,\mathrm{d}t \qquad (6.60c)$$

$$D_k = D[(k+1)T_s, kT_s] \qquad (6.60d)$$

The remaining matrices are as defined by eqn. 6.17.

The optimal control given by eqn. 6.57 is based upon the unrealistic assumption that all the state variables are measured. When this is not the case similar reasoning to that presented previously in this section leads us to believe that eqn. 6.57 would be replaced by

$$u_k^0 = -F_k\hat{x}_{k/k-1} - G_k d_k \tag{6.58}$$

where the first term corresponds to eqn. 6.37. Fig. 6.9 shows the stochastic setpoint control when some of the disturbances, but not all of the states, are measured.

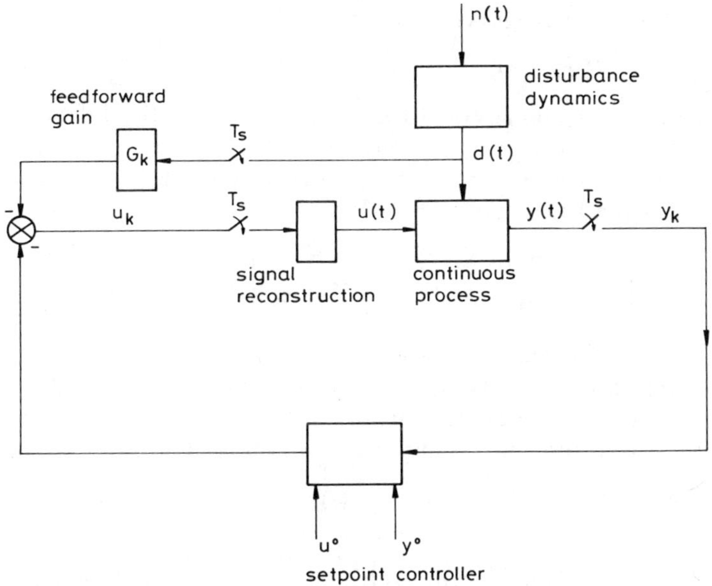

Fig. 6.9 *Setpoint control with measurable disturbances*

A *feedforward link* (Kwakernaak and Sivan, 1972) is the immediate consequence of permitting measurable disturbances. It is interesting to note how the optimal solution to the problem confirms the experience of many process control engineers who have successfully applied feedforward control on a more-or-less heuristic basis. Often even a crude disturbance model can result in much improved responses. It is very important to appreciate, however, that the feedforward link must always be used in conjunction with a feedback link, and never alone. Since the computation of $F_k\hat{x}_{k/k-1}$ is completely independent of the properties of the disturbances, the process control engineer may choose to start with only feedback (i.e. the stochastic setpoint control of Section 6.4) and later add feedforward links if the controlled loop performance is unsatisfactory. The stability of the controlled process is unaffected by the presence of the feedforward link.

The final topic of this section is the performance of the stochastic setpoint controller when subject to a change in the setpoint. From time to time changes in product demand, feedstocks, market prices etc. will necessitate a new (static) process optimisation, and new steady-state setpoints (u^0 and y^0) will be generated. We know that the regulator will function well at the new setpoint (it is independent of the value of the setpoint, and moreover it is robust) *provided* the linear process model is still a reasonable representation of the actual process dynamics. Thus our concern is really only how the changeover from one setpoint to another is handled by the regulator. Rather than consider just one changeover it is expedient from the viewpoint of the regulator design to model a number of abrupt changeovers by means of

$$r_{k+1} = H_k' r_k + n_k \tag{6.59}$$

Here, r_k is an $(n \times 1)$ dimensional vector that represents the difference between the new and the old setpoint (the constant vector x^0). The variable n_k is again zero-mean white noise. Augmenting r_k, which we consider as a new set of state variables, with the EDS state gives

$$\begin{bmatrix} x_{k+1} \\ r_{k+1} \end{bmatrix} = \begin{bmatrix} \Phi_k & 0 \\ 0 & H_k' \end{bmatrix} \begin{bmatrix} x_k \\ r_k \end{bmatrix} + \begin{bmatrix} \Gamma_k \\ 0 \end{bmatrix} u_k + \begin{bmatrix} w_k \\ n_k \end{bmatrix} \tag{6.60}$$

as the new model for regulator design.

The cost criterion must be appropriately modified: in view of the wish to minimise deviations of the actual state of the process (*not* the usual perturbation variable $x(t)$) from the new setpoint value an integral criterion of the form

$$\int_0^\infty \{(x(t) - r(t))^T Q(t)(x(t) - r(t)) + u^T(t)R(t)u(t)\}\, dt \tag{6.61}$$

$$= \int_0^\infty \left\{ [x^T(t) r^T(t)] Q(t) \begin{bmatrix} I & -I \\ -I & I \end{bmatrix} \begin{bmatrix} x(t) \\ r(t) \end{bmatrix} + u^T(t) R(t) u(t) \right\} dt \tag{6.62}$$

is applicable. The new optimal sampled-data regulator problem of minimising eqn. 6.62 subject to eqn. 6.60 is solved using the results of the previous two sections. We find that optimal control is given by

$$u_k^0 = -F_k x_k - G_k' r_k \tag{6.63}$$

where F_k is given, as when $r_k = 0$, by eqns. 6.15 and 6.16 while

$$G_k' = \tilde{R}_k^{-1} \bar{M}_k^T + (\tilde{R}_k + \Gamma_k^T \Sigma_k \Gamma_k)^{-1} \Gamma_k^T (\Sigma_k \bar{S}_k + P_k H_k') \tag{6.64}$$

Here, Σ_k is given by eqn. 6.16 while P_k is the solution of the equation

$$\begin{aligned} P_k = {} & S_k^T(\Sigma_k - \Sigma_k \Gamma_k (\tilde{R}_k + \Gamma_k^T \Sigma_k \Gamma_k)^{-1} \Gamma_k^T \Sigma_k) \bar{S}_k \\ & + S_k^T (P_k - \Sigma_k \Gamma_k (\tilde{R}_k + \Gamma_k^T \Sigma_k \Gamma_k)^{-1} \Gamma_k^T P_k) H_k' \\ & + \bar{N}_k - \tilde{M}_k \tilde{R}_k^{-1} \bar{M}_k^T \end{aligned} \tag{6.65}$$

The various newly-defined matrices are given by

$$\bar{M}_k \triangleq -\int_{kT_s}^{(k+1)T_s} H'^{\mathrm{T}}(t, kT_s)Q(t)\Gamma(t, kT_s)\,\mathrm{d}t \tag{6.66a}$$

$$\bar{S}_k \triangleq -\Gamma_k\tilde{R}_k^{-1}\bar{M}_k^{\mathrm{T}} \tag{6.66b}$$

$$\bar{N}_k \triangleq -\int_{kT_s}^{(k+1)T_s} \Phi^{\mathrm{T}}(t, kT_s)Q(t)H'(t, kT_s)\,\mathrm{d}t \tag{6.66c}$$

$$H_k' = H'[(k+1)T_s, kT_s] \tag{6.66d}$$

and the remaining matrices are as given by eqn. 6.17.

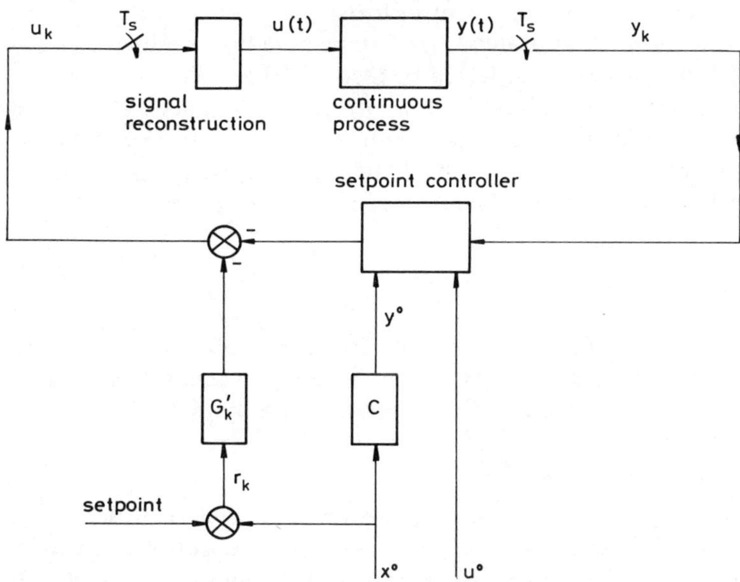

Fig. 6.10 *Setpoint control with changing setpoint*

When not all the state variables are measured, eqn. 6.63 should be replaced by

$$u_k^0 = -F_k\hat{x}_{k/k-1} - G_k'r_k \tag{6.67}$$

where $\hat{x}_{k/k-1}$ is determined by the KF. Fig. 6.10 shows the stochastic setpoint controller configuration to meet changes in setpoints. A closer examination of eqns. 6.63 to 6.65 reveals that the new controller may be computed using the same formulae as were previously used for determining the setpoint controller for measurable disturbances. Therefore, just as with that controller, the process control engineer may choose to initially implement only the state feedback link, i.e. $-F_k\hat{x}_{k/k-1}$ since this is completely independent of the dynamics of the changing setpoint.

6.6 S.i.s.o. setpoint control

The LQG setpoint controller assumes that the process dynamics are well known and predictable. It also tacitly assumes that the process will always be operated around one, given, setpoint which has been optimally calculated. In fact, the process model parameters are not known exactly and can change with time; the sampled-data setpoint regulator (although robust) then will no longer function optimally. It would be desirable to tune, online, the regulator, but the complexity of the regulator prohibits this: except, that is, if the LQG regulator could be decomposed into a number of s.i.s.o. controllers, each possibly accommodated in a microprocessor. Techniques for tuning s.i.s.o. controllers are well documented in many process control texts.

With this motivation in mind, we outline here how optimal s.i.s.o. setpoint controllers can be designed. It will be assumed that there are as many control signals as measured variables, i.e. $r = m$, and for simplicity all matrices in the process model description are taken to be time-invariant. Denoting the steady-state Kalman Filter gain matrix by K and the optimal sampled-data regulator gain matrix by F the optimal control can be expressed by

$$u_k = -F\hat{x}_k \tag{6.68}$$

$$\hat{x}_k = (S - KC - \Gamma F)\hat{x}_{k-1} + Ky_{k-1} \tag{6.69}$$

If the expressions for u_1, u_2, u_3 etc. are written down for the above two equations, it soon becomes apparent that a necessary and sufficient condition to ensure that $u_k^{(i)}$, $i = 1, 2, \ldots, m$ is determined only by

$$(y_{k-1}^{(i)}, y_{k-2}^{(i)}, \ldots, y_0^{(i)}\}$$

is that $F(S - KC - \Gamma F)^k K$ is a diagonal matrix for $k = 0, 1, 2, \ldots$. If this condition is fulfilled the optimal regulator may be constructed as m s.i.s.o. controllers. Obviously, FK must be made to be diagonal, and at the same time F and K so chosen that $(S - KC - \Gamma K)$ is αI where α is some scalar constant, or KSF where S is a diagonal matrix, for example.

It is extremely unlikely that the choice of Q, R, W or V that the process control engineer may decide on for the controller design will result in the above condition being fulfilled. We therefore adopt a different approach to the problem. First a number of possible F and K matrices are selected which satisfy the condition that $F(S - KC - \Gamma F)^k K$ is a diagonal matrix for each $k = 0, 1, \ldots$. Secondly, each pair of F and K selected is assumed to be the optimal feedback and Kalman gain matrices associated with an optimal control problem; the cost of using this pair is given by the minimum value of the cost criterion. Lastly, on the basis of the cost, the most favourable pair of F and K are chosen and used to implement the m s.i.s.o. controllers.

6.7 Actuator and sensor selection

It is rarely economically justifiable to base either actuator or sensor selection upon a detailed performance benefit/cost analysis investigation. Sometimes, however, a degree of freedom exists as to which of say two equally priced sensors should be employed for control purposes, each sensor measuring a different process condition (e.g. temperature, pressure, level etc.).

The approach to the design of s.i.s.o. controllers, outlined in the previous section, can be adapted to partially answer the sensor or actuator selection problem. It is a straightforward matter to compute the minimum cost associated with the use of a particular actuator configuration (Γ_k) or sensor configuration (C_k). The costs of the performance enhancement of various configurations may then be compared.

To reduce the computational burden there have been attempts to base selection upon some measure of the sensitivity of the reachability or observability Gramian to changes in Γ_k or C_k. However, it would be premature to counsel their use before results on the examination of more practical situations have been reported.

References

ANDERSON, B. D. O. and MOORE, J. B. (1981): 'Detectability and stabilizability of time-varying discrete-time linear systems', *SIAM J. Control & Opt.*, **19**, 1, pp. 20–32

ANDERSON, B. D. O. and MOORE, J. B. (1971): 'Linear optimal control' (Prentice-Hall)

ATHANS, M. (1972): 'The discrete-time linear-quadratic-gaussian stochastic control problem', *Annals of Econ. and Soc. Measurement*, **2**, pp. 449–491

BEVERIDGE, G. S. G. and SCHLECHTER, R. S. (1970): 'Optimization: theory and practice' (McGraw-Hill)

BOX, M. J. (1965): 'A new method of constrained optimization and a comparison with other methods', *Computer J.*, **8**, pp. 42–52

DOYLE, J. C. (1978): 'Guaranteed margins for LQG regulators', *IEEE Trans. Auto. Control*, **AC-23**, 4, pp. 756–757

GRIMBLE, M. J. (1983): 'Recent trends in the design of multivariable control systems by optimal methods'. Preprints Ist IASTED Int. Symp. Appl. Control and Ident., Copenhagen, I-1–I-9

GUIN, J. A. (1968): 'Modification of the Complex Method of constrained optima', *Computer J.*, **10**, pp. 416–417

HALYO, N. and CAGLAYAN, A. K. (1976): 'A separation theorem for the stochastic sampled-data LQG problem', *Int. J. Control*, **23**, 2, pp. 237–244

JOHNSON, A. (1983): 'Duality of the discrete-time optimal filter and regulator problems', *IEE Proc.*, **130**, pp. 173–174

JOSEPH, P. D. and TOU, J. J. (1961): 'On linear control theory', *Trans. AIEE*, **80**, 2, pp. 193–196

KALMAN, R. E. (1960): 'Contributions to the theory of optimal control', *Bol. Soc. Mat. Mexicana*, **5**, pp. 102–119

DE KONING, W. L. (1980): 'Equivalent discrete optimal control problem for randomly sampled digital control systems', *Int. J. Sys. Sci.*, **11**, 7, pp. 841–850

DE KONING, W. L. (1982): 'Infinite horizon optimal control of linear discrete time systems with stochastic parameters', *Automatica*, **18**, 4, pp. 443–453

DE KONING, W. L. (1984): 'Stationary optimal control and estimation of stochastically sampled continuous-time systems' (to be published)

KWAKERENAAK, H. and SIVAN, R. (1972): 'Linear optimal control systems' (John Wiley)

LEVIS, A. H., SCHLUETER, R. A. and ATHANS, M. (1971): 'On the behaviour of optimal linear sampled-data regulators', *Int. J. Control*, **13**, 2, pp. 343–361

MEIER, L., LARSEN, R. E. and THETHER, A. J. (1971): 'Dynamic programming for stochastic control of discrete systems', *IEEE Trans. Auto. Control*, **AC-16**, 6, pp. 767–775

ROSENBROCK, H. H. and STOREY, C. (1970): 'Mathematics of dynamical systems' (Nelson)

SAFONOV, M. G. and ATHANS, M. (1977): 'Gain and phase margin for multiloop LQG regulators', *IEEE Trans. Auto. Control*, **AC-22**, 2, pp. 173–179

SAGE, A. P. and WHITE, C. C. (1977): 'Optimum systems control' (Prentice-Hall)

WONG, P. K., STEIN, G. and ATHANS, M. (1978): 'Structural reliability and robustness properties of optimal linear-quadratic multivariable regulators', Proc. 7th IFAC World Congress, Helsinki, pp. 1797–1805

Index